Topological Semimetals

David J. Fisher

Published by **Materials Research Forum LLC**
Millersville, PA 17551, USA

Published as part of the book series
Materials Research Foundations
Volume 48 (2019)
ISSN 2471-8890 (Print)
ISSN 2471-8904 (Online)

Print ISBN 978-1-64490-014-7
ePDF ISBN 978-1-64490-015-4

Distributed worldwide by

Materials Research Forum LLC
105 Springdale Lane
Millersville, PA 17551
USA
http://www.mrforum.com

Printed in the United States of America
10 9 8 7 6 5 4 3 2 1

Table of Contents

Materials Research Forum LLC
doi: http://dx.doi.org/10.21741/9781644900154

Introduction to Topological Semimetals

During the past few decades there have been three major upheavals in what one might call the solid-state physics end of materials science. One was the discovery of quasicrystals; all the more surprising because it did not involve the creation of new materials, but merely involved looking at well-known and fairly mundane ones in a new light.

Another was the discovery of unsuspected properties being exhibited by materials which were better known for a different property; thus Heusler alloys, long known for their somewhat anomalous magnetic behaviour (being ferromagnetic in spite of containing no ferromagnetic elements) are now sought for their superconducting, semiconducting and indeed ... semimetal behaviours.

The third discovery has been the exciting properties which result from preparing the same material in an unfamiliar form, such as a glass or an atomically thin film; the prime example of the latter case being carbon when prepared in the form of graphene.

In the case of quasicrystals, it was a matter of looking at a relatively mundane alloy, going against the routinely accepted dogma that 5-fold crystalline symmetry was impossible, and seeing that alloy in a different light. The most recent and exciting trend, reviewed here, recalls the quasicrystal case in that it involves seeing old materials in a new light. But rather than involving a new insight into the crystallography of the arrangement of the material's atoms, the new insight concerns the topology of the quantum-mechanical behaviour of its electrons. The relatively recent discovery of topological phases in the electronic states of solids has revealed an ubiquitous influence of topology on the properties of various materials. Many well-known materials have turned out to be topological materials, suddenly widening the possibilities in material science.

The compounds, Na_3Bi and Cd_3As_2 were predicted theoretically to be Dirac semimetals in 2012 and 2013, respectively. Experimental verification of that, in 2014, launched the initial intense study of topological semimetals. The later theoretical prediction of non-magnetic Weyl semimetals in the TaAs family then triggered further experimentation. The numbers of research papers on this subject have been increasing geometrically during the past few years, and so it is perhaps too soon to attempt to summarise its progress. On the other hand, there is already a bewildering list of materials which fall, or may not fall, into this class and so it is nevertheless a good moment at which to introduce them to a wider audience.

Put simply, topological semimetals are materials in which the conduction and valence bands touch or cross one another, and the crossings are protected by topological constraints; protection here meaning that the linear energy band dispersion is preserved as long as the system symmetry is unbroken. In contrast to ordinary metals, whose Fermi surfaces are two-dimensional, topological semimetals can exhibit protected one-dimensional Fermi lines or zero-dimensional Fermi points, which arise due to an intricate interplay between the symmetry and topology of the electronic wave functions. Their unique physical properties have opened up a wide range of possible applications in low-power spintronics, in opto-electronics, in quantum computing and in the environmentally important field of the 'harvesting' of energies which would otherwise go to waste immediately as heat. Among the practical possibilities offered by these materials, it is noted that one route to fabricating superconducting microstructures in the non-superconducting Weyl semimetal, NbAs, is simple ion bombardment[1]: the marked difference in the surface binding energies of niobium and arsenic naturally leads to enrichment of the former at the surface, thus forming a superconducting surface layer with a critical temperature of 3.5K. Because it is formed from the target itself, the perfect contact between the superconductor and the bulk may allow effective gapping of the Weyl nodes in the bulk, due to the proximity effect. Simple ion bombardment could thus become an industrial-scale means of fabricating topological quantum devices from mono-arsenides.

Meanwhile Weyl semimetals such as NbAs and TaAs exhibit a marked reactivity with simple molecules, such as those of oxygen, carbon monoxide and water, given that there are several active sites available for surface reactions such as adsorption, decomposition, formation of reaction products and recombination of decomposition products. When various chemical species are adsorbed onto a Weyl semimetal, strong lateral interactions occur between the co-adsorbed species leading, for example, to CO-promoted water decomposition at room temperature: the resultant -OH groups reacting with CO to form HCOO, an intermediate species in water–gas shift reactions. These facts clearly demonstrate that Weyl semimetals have a place in catalysis[2], whereas their poor ambient stability due to rapid surface oxidation is a handicap in electronic applications. The identification of highly efficient low-cost catalysts is one of the main quests of catalytic chemistry, and the main strategy is currently to increase the number and activity of local catalytic sites such as the edges of molybdenum disulfides for the purposes of hydrogen evolution. An alternative principle[3] is to go beyond local site-optimization by using topological electronic states to stimulate catalytic activity. In the case of hydrogen evolution reactions, excellent catalysts have been identified among the transition-metal monopnictides, and the topological Weyl semimetals: NbP, TaP, NbAs and TaAs. The combination of stable topological surface states and a high room-temperature carrier mobility, both ensured by the bulk Dirac bands of the Weyl semimetal, is tailor-made for high-activity hydrogen evolution catalyst design. When compared with their rival, graphene-based composites, where graphene offers only high mobility and no active catalytic sites, they are clearly superior.

It has been shown[4] that a Veselago lens, for potential use as the tip of a scanning tunnelling microscope, can be made from Weyl semimetals such as NbAs and NbP. The ballistic nature of Weyl-fermion transport within the semimetal tip, combined with the ideal focusing ability of Weyl-fermions by a Veselago lens on the tip surface, could create a very narrow electron beam going from the tip to the specimen surface. By using a Weyl semimetal probe-tip, the resolution of a scanning tunnelling microscope could be improved sufficiently to image individual electron orbitals or chemical bonds. The current-voltage characteristics of devices based upon Weyl semimetals have been predicted by using the Landauer formalism. The potential step and barrier were reconsidered for three-dimensional Weyl semimetals, by analogy to the two-dimensional material, graphene. Applying pressure to a Weyl semimetal, which lacks a center of spatial inversion, permits the modelling of matter under extreme conditions; such as those in the vicinity of a black hole. It was further noted that Cd_3As_2 and Na_3Bi, with an asymmetry in their Dirac cones, could be modelled by using a scaling factor. This factor then created additional no-propagation regions and condensed the appearance of resonances. It was predicted that, under external pressure, a topological phase transition could occur in Weyl semimetals in which the electron transport changed in character and became anisotropic; where a hyperbolic Dirac phase occurred with strong light absorption and photocurrent generation.

The orbital degree of freedom is fundamentally important in understanding the less usual solid-state material properties. Thus, high orbitals in optical lattices can be used to construct quantum analogies, for exotic models of topological semimetals, which can involve many-body states such as Bose-Einstein condensates. The concept of the topological semimetal can be applied[5] to photonic systems and to some metamaterials. Photonic-crystal versions of Dirac degeneracies are protected by various space symmetries, with Bloch modes spanning the spin and orbital sub-spaces. Dirac points can also be theoretically created in effective media via the intrinsic degrees of freedom of electromagnetism. A pair of spin-polarized Fermi-arc surface state analogues can be observed at the interface between air and Dirac metamaterials. Eigen-reflection fields exhibit a decoupling from a Dirac point to two Weyl points. There is also a topological correlation between a Dirac point and the vortex or vector beams of classical photonics.

The introduction of orbital degrees of freedom into chequerboard and hexagonal optical lattices some years ago promised new means for predicting novel quantum states of matter that then had no analogues in solid-state electronics. An exotic topological semimetal was expected to exist as a parity-protected gap-less state in the orbital bands of a two-dimensional fermionic optical lattice; that quantum state being characterized by a parabolic band-degeneracy point having a 2π Berry flux. This was very different to the π flux of Dirac points present in graphene. The appearance of this topological liquid was common to all lattices of D_4 point-group symmetry, provided that orbitals of opposite parity hybridized strongly with each other and that the band degeneracy was protected by an odd parity. In the presence of interparticle repulsive interactions, the system underwent a phase transition to a topological insulator which was expected to exhibit detectable chiral gap-less domain-wall modes[6]. A similar study of a two-dimensional fermionic square lattice which permitted the existence of a two-dimensional Weyl semimetal, an anomalous quantum Hall effect and a 2π-flux topological semimetal within various parameter ranges showed[7] that the band degenerate points of the Weyl and 2π-flux topological semimetals were protected by two distinct hidden symmetries. Both of these corresponded to anti-unitary composite operations. When the latter symmetries were broken, a gap opened up between the conduction and valence bands and made the system insulating. The degenerate point at the boundary between a quantum anomalous Hall, and trivial band, insulator was similarly protected by a hidden symmetry. A quantum anomalous Hall effect occurred for some parameter choices.

The detection of a quantized version of the anomalous Hall effect in thin films of a magnetically-doped topological insulator completed the 'hat-trick' of such effects; the others being the quantum Hall effect and the quantum spin Hall effect. Given that the intrinsic anomalous Hall effect is related to the Berry curvature and to the U(1) gauge field in momentum space, this establishes a link between the quantum anomalous Hall effect and the topological properties of electronic structures as characterized by the Chern number. When the time-reversal symmetry is broken by applying a magnetic field, a quantum anomalous Hall effect system exhibits a dissipation-free charge current at edges. This resembles the quantum Hall effect, although an external magnetic field is there necessary. The quantum anomalous Hall effect and the related Chern insulators are closely related to other topological states such as insulators and semimetals.

One theory for the anomalous Hall effect in a topological Weyl superconductor having broken time-reversal symmetry was to consider a ferromagnetic Weyl metal possessing two Weyl nodes, of opposite chirality, near to the Fermi energy[8]. In the presence of inversion symmetry, the metal then underwent weak-coupling Bardeen-Cooper-Schrieffer instability, with a pairing of the parity-related eigenstates. Such a superconductor was topologically non-trivial, with Majorana surface states coexisting with the Fermi arcs of the normal Weyl metal, due to the non-zero topological charge carried by the Weyl nodes. Under certain conditions, the anomalous Hall conductivity of such a superconducting Weyl metal was predicted to coincide with that of a non-superconducting one, in spite of the superconductor's non-conservation of charge. This behaviour was attributed to conservation of the chiral charge.

On-resonant light has been suggested to create a Floquet topological state in an ordinary band insulator. It has further been suggested that a Floquet Weyl semimetallic state might be created in three-dimensional non-magnetic or magnetic topological insulators by applying off-resonant light[9]. Virtual-photon phenomena are important in renormalizing the Dirac mass and producing a topological semimetal having a vanishing gap at Weyl points.

Further analogous topological effects can occur in bosonic systems when excited by finite-frequency probes. One approach likens the zero-frequency excitations of mechanical systems to the topological zero modes of fermions, such that mechanical models of topological semimetals can be constructed. The resultant gap-less bulk modes are physically different to the familiar acoustic Goldstone phonons, and appear even in the absence of a continuous translation invariance. The zero-frequency phonon modes also exhibit adjustable momenta and are protected topologically, provided that the lattice coordination remains unchanged[10]. Because these fundamental topological principles apply to any type of wave under periodic boundary conditions, Weyl point analogues are expected to exist in classical wave systems. An acoustic Weyl phononic crystal could thus be created[11] by breaking space-inversion symmetry via a combination of slanted acoustic waveguides. These acoustic Weyl points could be characterized by means of angle-resolved transmission measurements. Analogous acoustic Fermi arcs could moreover be observed by scanning the distribution of surface waves. One-way acoustic transport was also observed, in which the surface waves could overcome step barriers without suffering any reflection.

As hinted above, the topological semimetals also provide an interesting and unexpected test-bed for examining some of the most fundamental current theories in physics and cosmology. Study of Weyl and Dirac topological materials, including topological semimetals, insulators, superfluids and superconductors, indeed offers a path towards the investigation of the topological quantum vacua of relativistic fields. By analogy, the symmetrical phase of the standard model, where neither electroweak nor chiral symmetry are broken, represents a topological semimetal[12]. Vacua of the standard model in phases having broken electroweak symmetry represent topological insulators of various types.

Reversal of the analogy helps to answer questions concerning the stability of the vacuum. These aspects, of what is an already complicated subject, will be only touched upon here.

Already bewildering enough is the proliferation of quantum-mechanical phenomena and an associated menagerie of quasi-particles which ultimately account for the useful and interesting properties of these materials, and particularly the possibility of tailoring those properties by understanding and exploiting the new discovery. Attention will be concentrated here on the observable properties however, and reference made only in passing to the often extremely subtle underlying quantum processes.

A good point of entry into the subject is the case of graphene (figure 1). This material is best known for its astonishing mechanical properties, but its equally interesting electronic properties and their quantum-mechanical origins put it into the class of topological semimetals. To those materials scientists who take an interest only in the mechanical properties of graphene, its other aspects seem to take on a somewhat 'Alice-in-Wonderland' quality. For example, graphene is often described as being a free-standing monolayer of graphite, and it is therefore initially puzzling to find some topological semimetals being called 'three-dimensional graphene'. One might well ask why three-dimensional graphene is not simply again called … graphite. The answer lies of course in all of the quantum-mechanical subtleties which are involved.

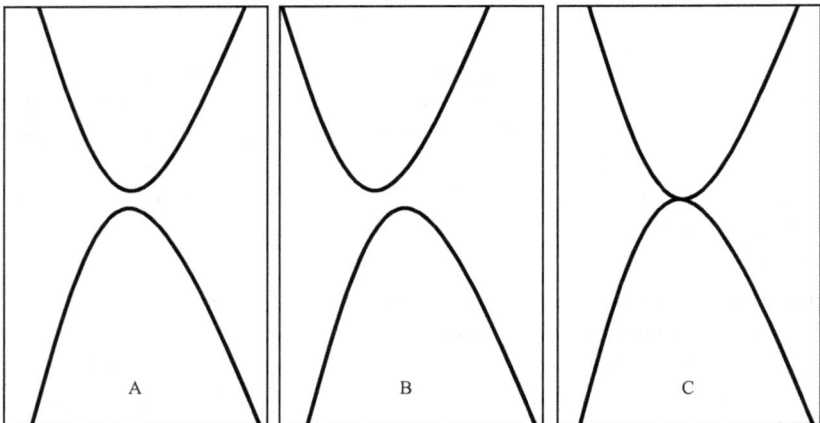

Figure 1. Various arrangements of valence and conduction bands
A: direct band-gap semiconductor (e.g. GaAs), B: indirect band-gap semiconductor (e.g. Si,) C: semimetal (e.g. graphene multilayer)

In fact, rhombohedral graphite does behave like a topological semimetal. It has flat surface sub-bands, and the bulk is semimetallic. A bulk-surface correspondence arises from an ABC-stacking of graphene layers. Bulk sub-bands in rhombohedral graphite can be viewed as being a three-dimensional Dirac-cone structure in which the Dirac-points form continuous lines that spiral in momentum-space. Changes in the gapped bulk sub-bands of n-layer ABC-stacked graphene which occur with increasing n, and their conversion into a three-dimensional Dirac-cone structure in the bulk when the bulk gap closes up at the Dirac-point spirals, has been examined[13] by using a non-perturbative effective Hamiltonian. It is found that the wavelength of the standing-wave function encompassing the stack of layers depends upon the in-plane Bloch momentum such that, in the bulk limit, coupling vanishes and the wavelength is thus irrelevant to the surface behaviour.

Three-dimensional topological semimetals can support band crossings along one-dimensional curves (nodal or Dirac lines) in momentum space, protected by structural symmetries and topology. In rhombohedrally (ABC) stacked honeycomb lattices supporting Dirac lines protected by time-reversal, inversion and spin rotation symmetries, there tends to exist a pair of nodal lines in momentum space which extend, through the entire Brillouin zone, in the stacking direction. These Dirac lines are topologically distinct from the usual Dirac lines, which form closed loops within the Brillouin zone. An energy-gap can then be opened only by merging pair-wise the Dirac lines going through the Brillouin zone so that they turn into closed loops within the zone, and then shrinking the loops to points[14]. This type of topological phase transition can occur in rhombohedrally stacked honeycomb lattices upon adjusting the ratio of the tunneling amplitudes in directions perpendicular to, and parallel to, the layers.

The entire energy spectrum of a square lattice, made up of superconducting quantum circuits, has been measured. The tunable gap-less band-structure which was typical of a topological semimetal was observed and, by tuning the parameters, it was possible to cause a quantum phase transition from a space-time semimetal - a form of topological semimetal - to an insulator by merging a pair of Dirac points. The time reversal and inversion symmetries could be broken while maintaining space-time symmetry[15]. In effect, a topological semimetal phase was created in the superconducting quantum circuit by imitating the momentum space using a modified parameter space.

The effect has been studied of a space-dependent Weyl-node separation, treated as a background axial-vector potential, upon the electromagnetic response and the energy spectrum of Weyl and Dirac semimetals[16]. This served to model the situation which can be created in the solid state by an inhomogeneous strain or a non-uniform magnetization.

A semi-classical approach showed that a resultant axial magnetic field was manifested by an increase in the conductivity, due to a chiral pseudomagnetic effect. Further analysis of the effect of the axial magnetic field upon the spectral properties revealed a pseudo-Landau level-structure for various magnetic-field spatial profiles. This showed that the Fermi-arc surface states of Weyl semimetals could be treated as pseudo-Landau levels which resulted from a magnetic field that was confined to the surface. Due to position-momentum locking, a bulk field could create pseudo-Landau levels which linked, in real space, the Fermi arcs on opposite surfaces. There were also equilibrium bound currents which were proportional to the field, but which averaged to zero over the sample. These could be regarded as being the analogues of bound currents in magnetic materials.

An investigation has been made of two types of multiple-Q magnetically ordered states, and of a topological phase transition between them in two dimensions. One was a co-planar but non-collinear double-Q state on a square lattice (a semimetal which accommodated massless Dirac electrons) and the other was a non-coplanar triple-Q state on a triangular lattice (a Chern insulator exhibiting an anomalous quantum Hall effect). The electronic structures of the two multiple-Q states could be analyzed in a unified manner by using the Kondo lattice model. This then suggested that a quantum phase transition occurred between the two states via a continuous change in lattice geometry between the square and triangular lattices[17]. Mean-field approximations to the ground state of the periodic Anderson model confirmed that a continuous topological phase transition occurred between the double-Q Dirac semimetal and the triple-Q Chern insulator on a square-to-triangular lattice.

Considering yet another graphene-related material with ABC-stacking, first-principles calculations predict[18] graphdiyne to be a nodal-line semimetal with Z_2 monopole charges forming a linking structure. As might be expected, there are many analogies between the local and global charges of topological semimetals, and certain features of singular vector fields. For example, the semimetal can be modelled as Euler chains and, from these, the surface Fermi-arc connectivity can be predicted. The analogies, and links to the topological invariants of insulators, can be understood by using exact geometrical sequences[19]. Carrying the analogies beyond known parallels then predicts the existence of semimetals in which the local charges might be constrained Atiyah–Dupont–Thomas invariants. One experimental clue to their actual existence would then be the observation of torsional Fermi-arcs.

A formula was developed for computing the second Stiefel-Whitney class based upon parity eigenvalues of inversion-invariant momenta and clarifying the quantized bulk magneto-electric response of nodal-line semimetals with Z_2 monopole charges under time-reversal symmetry-breaking perturbation. Unlike PT-symmetry nodal lines which are protected only by the π Berry-phase, where only a single nodal line can exist, nodal lines with Z_2 monopole charges exist in pairs. A pair of nodal lines with Z_2 monopole charges is created by a double band-inversion process, and the resultant nodal lines are always linked by another nodal line which forms between the two top-most occupied bands. Delving into abstract topology shows that the linking structure and Z_2 monopole charge are aspects, of a non-trivial band topology of the second Stiefel-Whitney class, which can be deduced from the Wilson loop spectrum. That class serves as a well-defined topological invariant of a PT-invariant two-dimensional insulator in the absence of the Berry phase. It can thus be surmised that pair-creation and the annihilation of nodal lines with Z_2 monopole charges can mediate topological phase transitions between a normal insulator and a three-dimensional weak Stiefel-Whitney insulator.

An apparent anachronism is to be found in the fact that the underlying quantum-mechanical phenomena which make topological semimetals so interesting were predicted to exist in graphene ... long before graphene *per se* was created. One finds, for example, Weyl speculating on fermionic aspects of electron propagation in solids nearly ninety years ago[20] and, some seventy years ago, the meaning of electron-mass in graphite monolayers was considered[21,22]; both paving the way for the recognition of the importance of spin and quasi-particles in 'two-dimensional' solid-state physics. 'Weyl' is appropriately now the name of one of the main classes of topological semimetals.

The connection will not be pursued very far here, but there is a close theoretical analogy between topological semimetal theory and the wider world of particle-physics, given that they both involve discussion of the same particles: fermions. The analogy is not perfect of course because relativistic effects are not so prominent a consideration in solid-state physics. Although Lorentz invariance is required in high-energy physics, it is not necessarily obeyed in condensed-matter physics. On the other hand, some posited species of fermions which cannot actually exist in particle-physics are nevertheless to be found in the solid state. The discovery of topological insulators for example encouraged an intensely competitive search for new topological semimetals, including Dirac semimetals, Dirac nodal-line semimetals and Weyl semimetals, because such materials host the massless fermions which are the condensed-matter embodiment of long-sought relativistic fermions in high-energy physics. Because of the absence of Lorentz-invariance considerations, Lorentz-violating type-II Weyl and Dirac fermions can be found in topological semimetals. Lorentz considerations can sometimes be of diagnostic interest: semimetals can have point-like Fermi surfaces in various spatial dimensions which occur naturally in the transition between a weak topological insulator and a trivial insulating phase. Such phases are constructed by layering strong topological insulator phases in the next-lower dimension. Their response is generally of a form which represents a source of Lorentz violation and can be deduced from the location of the nodes in momentum space and from the helicities/chiralities of those nodes[23].

An antiferromagnetic semimetal has been identified[24] as perhaps hosting three-dimensional symmetry-protected Dirac fermions. Reorientation of the Néel vector can break the underlying symmetry and open up a gap in the quasi-particle spectrum; thus provoking a semimetal-insulator transition. It is expected that such a transition can be controlled by manipulating the chemical potential of the material. Analytical and numerical studies of the thermodynamic potential of the model Hamiltonian indicate that a gapped spectrum is preferred when the chemical potential is located at the Dirac point. As the potential wanders away from that point, the system can possibly undergo a transition from the gapped to gap-less phase.

Fermions - elementary particles such as electrons - are classified as being Dirac, Majorana or Weyl. Majorana and Weyl fermions had not been observed experimentally until the discovery of condensed matter systems such as topological superconductors and semimetals, in which they arise as low-energy excitations. In quantum field theory, Lorentz invariance leads to the three types of fermion. The existence of Weyl and Majorana fermions as elementary particles is still controversial in high-energy physics. All three types of fermion have meanwhile been proposed to exist as low-energy, long-wavelength quasi-particle excitations in condensed matter. The existence of Dirac and Weyl fermions in condensed matter has been confirmed experimentally, and there is some experimental support for that of Majorana fermions. As previously noted, Weyl proposed in 1929 that the massless solution of the Dirac equation represents a pair of a new type of particle: the Weyl fermion. The rapid advances in the study of topological insulators and topological semimetals have provided a means for generating Weyl fermions in condensed matter: when two non-degenerate bands in three-dimensional momentum space cross at points, so-called Weyl nodes, near to the Fermi energy, the low-energy excitations behave just like Weyl fermions. The chirality of Weyl fermions can be directly determined by measuring the photocurrent that results from exposure to circularly polarized mid-infrared light[25]. The magnitude of the photocurrent is governed by the chiralities of the fermions and the photons. Those two chiralities are analogous to the two valleys of two-dimensional materials, and lend a new degree of freedom to a three-dimensional crystal.

Before leaving this topic of the analogy between condensed-matter quasi-particles and their analogues in fundamental high-energy physics, it is instructive to see just how far the analogy can be pushed. It turns out that effective gravity and gauge fields are

Topological Semimetals   Materials Research Forum LLC
Materials Research Foundations **48** (2019)   doi: http://dx.doi.org/10.21741/9781644900154

emergent properties which can be associated with the low-energy quasi-particles of topological semimetals. Using Dirac semimetals as examples, it has been shown[26] that an applied lattice strain can simulate astrophysical phenomena such as warped spacetime, the black-hole event horizon, Hawking radiation and gravitational lensing.

An extensive cataloguing of possible space groups has shown[27] that gap-closing in three-dimensional inversion-asymmetry crystals always leads to the creation of a Weyl or nodal-line semimetal. The space group, and the wave vector at gap-closing, together govern which alternative occurs. Following closure of the gap, the gap-closing points or lines lie in the wave vector space. Contrary to the case of inversion-symmetry systems, no insulator-insulator transition ever occurs.

In condensed matter, the existence of fermions in crystals is limited by the symmetries of the 230 crystal space groups rather than by Lorentz invariance, but this in turn raises the possibility of other types of fermionic excitation that have no equivalent in high-energy physics. That form of electron filling which is compatible with band insulator behaviour has been calculated[28,29] for all of the space-groups, assuming the presence of non-interacting electrons which exhibit time-reversal symmetry. The results furnished simple criteria for identifying topological semimetals and for judging the stability of the resultant nodal Fermi surfaces.

A complete classification of all of the possible non-symmorphic band degeneracies in hexagonal materials with strong spin-orbit coupling has been developed[30] on the basis of the algebraic relationships obeyed by the symmetry operators, and the compatibility between irreducible representations at various high-symmetry points of the Brillouin zone. It includes band-crossings which are protected by conventional non-symmorphic symmetries where partial translation lies within the invariant space of the mirror/rotation symmetry. It further includes band-crossings which are protected by off-center mirror/rotation symmetries in which the partial translation is orthogonal to the invariant space.

Non-symmorphic symmetries, such as screws and glides, produce electron band touchings, obstruct the formation of band insulators and lead to metals or nodal semimetals even when the number of electrons in the unit cell is an even integer. With regard to the properties of electrons in magnetically ordered crystals, which are of interest to the creation of, for example, topological phases such as magnetic Weyl semimetals, it is worthwhile mentioning an even more heroic systematic study. This involved[31] using a representation of allowed band structures to obtain a description of the basic properties of free electrons in all of the *magnetic* space groups. Unfortunately, there were 1651 magnetic space groups to be considered in three dimensions, as well as 528 magnetic layer groups which arise in two dimensions.

A more traditional search method was also applied to the identification of relevant behaviour among the 18-valence electron family of ABX compounds. It was noted that, at that time, only 83 of the 483 possible compounds had been prepared. First-principles thermodynamic studies of the theoretical stability of the remaining 400 compounds then predicted that just 54 should be stable. Of those, 15 were then prepared[32] and the predicted crystal structures were all confirmed. The still-remaining unexamined compounds were predicted to include possible transparent conductors, thermoelectric materials and topological semimetals.

A further proposed addition to the menagerie of topological semimetals is the Hopf semimetal[33]. This is characterized by a number of the form, pq, and the Fermi surface geometry is then dictated by a Hopf map involving linked loops such that the surface becomes a torus link of some type. The Hopf link, for example, is characterized by p = 1 and q = 1 while a so-called Solomon knot is characterized by p = 2 and q = 1. Further examples are the double-Hopf link with p = 2 and q = 2, and the double-trefoil knot with p = 3 and q = 2. As well as integers, p and q can also be half-integers or arbitrary rational numbers.

All of the possible types of intersecting nodal ring which can exist in layered semiconductors having Amm2 and Cmmm space groups have been classified[34], and a critical condition has been established for the existence of intersecting nodal rings in symmorphic crystals. Certain honeycomb structures have been suggested to be intersecting nodal-ring topological semimetals, including some having layered and so-called hidden layered structures. In these structures, transitions between three types of intersecting nodal ring can be driven by an external strain. It is found that the resultant surface states and Landau-level structures can be more complicated than those which arise from simple nodal loops. A two-band model can describe multiple closed nodal-loop semimetals, and all such known cases (nodal-net, nodal-chain, Hopf-link, etc.) can be treated by using the same framework[35]. The topologically different bulk states can be compared on the basis of a double nodal-loop model, including the corresponding drumhead surface states. There is also a connection with Hopf insulators. Regardless of the direction of the magnetic field, the Hopf-link state is characterized by the existence of a quadruply degenerate zero-energy Landau band.

A new scheme[36] for the identification of non-magnetic topological semimetals uses, as input, only inversion- and rotation-symmetry eigenvalues of the valence bands at high-symmetry points of the Brillouin zone. It returns the type (line or point), topological charge, number and configuration of all of the existing stable topological band-crossings. All of the latter are located at generic momenta in systems possessing time-reversal symmetry and negligible spin-orbital coupling.

Study of topologically non-trivial semimetals by using the six-band Kane model can demonstrate the existence of surface states by calculating the local density of states on the material's surface. In the strain-free condition, the surface states are divided into a part which is in the direct gap and a part, including the crossing-point of the surface-state Dirac cone, which is in the valence band[37]. Application of a uniaxial stress creates an insulating band-gap, and moves the crossing point-from the valence band and into the band-gap, to make the system a topological insulator while disorder markedly increases the spin-Hall effect in the valence band of thin films.

Returning now to the graphene connection, the search for carbon-based materials exhibiting topological behaviours has become an important new field of research. First-principles calculations and tight-binding modeling, for example have been used[38] to identify Weyl semimetals which are based upon three-dimensional graphene networks. Their band structures contain two flat Weyl surfaces, in the Brillouin zone, which straddle the Fermi level and resist external straining. When the networks are broken up, the resultant pieces remain semimetallic, with Weyl lines and points existing at the Fermi surfaces of slices and nanowires, respectively. Between the Weyl lines, flat surface bands emerge which can exhibit a marked magnetism. This notable structural stability can be attributed to a bulk topological invariance which is imposed by the sub-lattice symmetry, and to the one-dimensional Weyl semimetal behavior of the zig-zag carbon chains. The stability of Dirac semimetals, such as graphene, in two spatial dimensions requires the presence of lattice symmetries. Weyl semimetals in three spatial dimensions are protected by band topology. In the bulk of topological band insulators, self-organized topologically protected semimetal material can exist along a grain boundary. These states exhibit a valley anomaly in two dimensions which affects edge spin transport. In three dimensions, they appear as graphene-like states which can exhibit an odd-integer quantum Hall effect[39].

The integer quantum Hall state can be regarded as being a prototype for topological matter, including the topological insulator. The topological side of the integer quantum Hall state is usually discussed in terms of the Thouless-Kohmoto-Nightingale-den Nijs formula, which links the Berry phase to the Hall conductivity. The non-trivial topology of the topological insulator arises from the often concealed presence of a Dirac monopole in Hamiltonian parameter space. Given its identical Dirac monopole structure, the Hamiltonian which describes the Rabi oscillation contains the seed of a topological insulator. The next step in research on topological matter is expected to be to examine interaction-induced topological phases such as the fractional Chern and topological insulators. The latter have been theoretically predicted on the basis of an analogy with fractional quantum Hall states.

The general mechanism, the hybridization of spinon modes bound to the grain boundary, suggests that topological semimetals can exist in any topological material where lattice dislocations bind localized topological modes. First-principles calculations predicted[40] the existence of a topological semimetal which exhibits nodal nets that comprise multiple interconnected nodal lines in the bulk and have two coupled drumhead-like flat bands around the Fermi level on its surface. It was proposed that this nodal-net semimetal state would exist in a graphene-network structure that might be constructed by inserting a benzene ring into each C-C bond in the body-centred tetragonal C_4 lattice, or by crystalline modification of (5,5) carbon nanotubes. The Weyl semimetal is a three-dimensional topological state of matter in which the conduction and valence energy bands touch at a finite number of nodes. The nodes always appear in pairs, in each pair the quasiparticles carry opposite chirality and linear dispersion, much like a three-dimensional analog of graphene. A topological quantum phase requires a finite momentum-space Berry curvature which can arise through breaking the inversion or the time-reversal symmetry so as to generate non-trivial topologically invariant quantities associated with the underlying energy band structure (e.g., a finite Chern number). In the case of conventional graphene or graphene-like two-dimensional systems with gap-less Dirac cones, symmetry-breaking renders the system insulating due to lifting of the degeneracy. It is of interest to design materials that simultaneously possess both a semimetal phase with gap-less bulk Dirac-like cones and a finite Berry curvature. A two-dimensional mechanical lattice system has been proposed[41] which exhibits those properties. An intrinsic valley Hall effect can then appear without impairing carrier mobility, as the quasi-particles remain massless. Upon confinement along the zig-zag edges, two different types of gap-less edge states having opposite edge polarizations can appear. One of these has a finite group velocity while the other has a zero group velocity. Thermal transport in most carbon allotropes is mediated by phonons, and the atomic bonding thus directly affects phonon transport. One allotropic phase is a topological semimetal with (like graphene) an sp^2 bonding network and a 16-atom body-centered orthorhombic unit cell. The other allotrope is constructed by replacing each atom in diamond by a carbon tetrahedron, and has an sp^3 bonding network (like diamond). By comparing density functional theory results, the cause of the differing lattice thermal conductivities of the allotropes was revealed[42] to arise at the atomic-bonding level, in spite of the similar hybridization.

Various types of topological semimetal which exhibit a bulk quadrupole moment have been predicted[43] on the basis of a model which involved a three-dimensional extension of a two-dimensional quadrupole topological insulator. One of the semimetals was predicted to have bulk nodes and gapped topological surfaces. Another type was gapped in the bulk, but featured a Dirac semimetal having an even number of nodes on one or more surfaces. The third type was gapped in the bulk but featured half of a Dirac semimetal on several surfaces. Due to the bulk quadrupole moment, each of the predicted semimetals generated mid-gap hinge states and a hinge charge, as well as surface polarization. The bulk quadrupole moments could be deduced from the momentum locations of bulk and surface nodes in the energy spectrum. In order to achieve this, it could be advantageous to consider nodes in the Wannier bands, rather than energy bands.

The possibility of a magnetic second-order topological insulator was explored[44] by using a three-dimensional model. It was noted that a previously proposed topological hinge insulator had surface states lying along the [001] direction in addition to hinge states. These surface states were treated by introducing magnetization and by obtaining a second-order topological insulator with only hinge states. The bulk topological number was the Z_2 index, protected by combined four-fold rotation and inversion symmetries. A study of two-dimensional magnetic second-order topological insulators showed that the corner states were stable in the presence of magnetization. A magnetic second-order topological semimetal was constructed by layering the two-dimensional magnetic second-order topological insulator, where hinge-arc states were stable in the presence of magnetization.

Dirac Semimetals

Dirac topological semimetals with accidental band touching between conduction and valence bands (figure 1) protected by time reversal and inversion symmetry are at the frontier of modern condensed matter research. The Dirac node is composed of two Weyl nodes of opposite chirality which overlap one another in momentum space (figure 2). The majority of known topological semimetals are non-magnetic and conserve time reversal symmetry.

As explained earlier, the Dirac semimetal can be regarded as being a three-dimensional version of graphene. Its existence has been theoretically and experimentally confirmed in the case of Na_3Bi and Cd_3As_2, where the bands cross near to the Fermi energy and form Dirac-points of four-fold degeneracy and with the band-crossing points being protected by crystalline symmetry. It will be useful to consider those materials in detail.

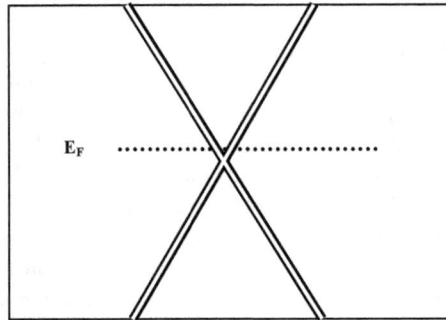

*Figure 2. The Dirac node comprises two Weyl nodes of
opposite chirality which overlap in momentum space*

Topological semimetals are noted for a linear energy band dispersion in the bulk state, and for topologically-protected surface states having arc-like Fermi features. Angle-resolved photo-emission spectroscopy has confirmed the existence of linear Dirac (Weyl) cones and Fermi arcs. The presence of a three-dimensional Dirac semimetal phase is also revealed by the observation of Shubnikov-de Haas oscillations, while the existence of a Weyl fermion-related chiral anomaly effect is signalled by the observation of negative magnetoresistance and thermal power suppression.

In topological semimetals which are exposed to weak magnetic fields, there is a competition between the positive magnetoresistivity which is produced by the weak anti-localization effect, and the negative magnetoresistivity which is related to non-trivial Berry curvature. The weak localization is assumed to be caused by intervalley and interaction effects and to occur in double-Weyl semimetals.

In disordered metals, weak localization and anti-localization transport phenomena arise from quantum interference. At low temperatures, they can lead to characteristic temperature and magnetic field dependences of the conductivity and thereby permit the system symmetry to be explored. In the case of topological materials, the quasiparticles are termed Dirac or Weyl fermions. In the case of the negative magnetoresistance caused by the non-trivial Berry curvature in topological semimetals, there is a dependence of the negative magnetoresistance upon the carrier density. In the quantum limit, and in strong magnetic fields, the magnetoconductivity depends upon the type and range of the disorder scattering potential. As noted elsewhere, a positive magnetoconductivity in a strong magnetic field may not be a definite indication of chiral anomaly. The magnetoconductivity can be linear in the quantum limit for half-filling and a long-range Gaussian scattering potential[45]. Some conductivity is observed at the Weyl nodes, even though the density of states vanishes there.

A connection has been found between the parity of the monopole charge in a topological semimetal, and quantum interference corrections to the conductivity, such that the parity of the monopole charge determined the sign of the quantum interference correction[46]. That is, odd and even parities led to weak anti-localization and weak localization, respectively. This was attributed to the Berry-phase difference between the time-reversed trajectories which circled the Fermi sphere that enclosed the monopole charge. The low-temperature weak-field magnetoconductivity was proportional to +B in the case of double-Weyl semimetals and proportional to -B in the case of single-Weyl semimetals.

The chiral anomaly is generally assumed to lead to positive magnetoconductivity or negative magnetoresistivity in strong and parallel fields. Some studies of Weyl and Dirac topological semimetals have nevertheless indicated the occurrence of negative magnetoconductivity in high fields. In the case of a strong magnetic field, applied along the line connecting two Weyl nodes, further work has shown that the conductivity along the field direction is governed by the Fermi velocity rather than by the Landau degeneracy. The identification of three situations, in which the high-field magnetoconductivity was negative, suggested that a high-field positive magnetoconductivity is not necessarily a definite indicator of the chiral anomaly[47].

When an external magnetic field is imposed, the degenerate Dirac point splits into two separate Weyl nodes in the magnetic field direction, transforming a Dirac semimetal into a Weyl semimetal. The chiral anomaly effect thus appears when the magnetic field is parallel to the electric field: a chiral charge current is driven from one Weyl node to the other, leading to an additional electric conductivity and the negative magnetoresistance.

The surface states near to the projection of a Dirac point can also be treated as being the superposition of a helicoid and anti-helicoid of differing chirality, but there are still crossings of the bulk states and of the surface states. If there is no additional symmetry-protection, such as band-crossing, hybridization of two Weyl nodes of opposite chirality can open up a gap. The linear energy-band dispersion would then disappear, together with the Fermi-arc. Yet more conditions, such as non-symmorphic symmetry, are therefore required by three-dimensional Dirac semimetals.

A three-dimensional Dirac semimetal has bulk Dirac cones in all three momentum directions and Fermi-arc like surface states, and can be converted into a Weyl semimetal by breaking time-reversal symmetry.

The topological semimetal state has been further generalized[48] to the concept of a higher-dimensional Dirac nodal sphere or pseudo Dirac nodal sphere state. The band-crossings here form a two-dimensional closed sphere at the Fermi level. It was demonstrated that two types of pseudo Dirac nodal sphere state were possible in real crystals, supported by different crystalline symmetries which involved a spherical backbone that consisted of multiple Dirac nodal lines and an approximate band-degeneracy between the Dirac nodal lines. Strained YH_3, HoH_3, TbH_3, NdH_3 and Si_3N_2 were predicted to exhibit the two types of state.

Weyl Semimetals

Beginning with the discovery of two-dimensional and three-dimensional topological insulators, the study of topological states has rapidly become one of the most intensively studied fields of condensed-matter physics. In topological insulators, the band inversion caused by strong spin-orbital coupling leads to opening of a band-gap in the entire Brillouin zone, whereas an additional crystal symmetry such as point-group and non-symmorphic symmetries sometimes prohibits gap-opening at/on specific points or a line in momentum space, giving rise to topological semimetals. Despite many theoretical predictions of topological insulators/semimetals associated with such crystal symmetries, the experimental realization is still relatively scarce.

The total number of Weyl nodes (figure 3) has to be even because the magnetic charge described by the Berry curvature has to be zero in a band structure. The minimal Weyl semimetal therefore possesses just a pair of Weyl nodes, and this can be achieved only in a time-reversed symmetry-breaking system; a situation usually imposed by magnetic order.

When the electrons in the low-energy region obey the Weyl equation, the semimetal is termed a Weyl semimetal. The conductance and valence bands here intersect at points in momentum-space termed Weyl nodes. Around these nodes, the low-energy behaviour is described by a three-dimensional two component term which is the dot-product of the Pauli matrix and the crystal momentum. This is in turn multiplied by a scalar factor which can take a +1 or -1 value, depending upon the chirality of the Weyl node.

The Fermi surface of topological semimetals ideally exhibits isolated band-crossing points or lines rather than a surface, as in the case of normal metals. The band-crossing points are a consequence of band-inversion and behave like Berry-flux monopoles; producing a quantized Berry flux on the Fermi surface that surrounds a nodal point or a Berry-phase along a loop that threads through the nodal line. The associated characteristic numbers can be used as topological invariants by which to characterize the band topology. This then leads to the identification of at least three species of non-trivial topological semimetal, based upon the degeneracy of the band-crossing points and their distribution within the Brillouin zone: the Weyl semimetal, the Dirac semimetal and the nodal-line semimetal. These all have band-crossing points due to band-inversion.

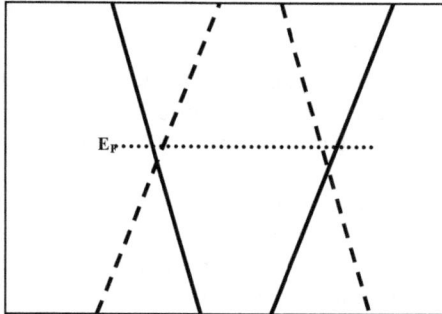

Figure 3. The Weyl semimetal band-crossings are isolated points with two-fold degeneracy, and are effectively monopoles in the Brillouin zone

Many materials have been suggested to host Weyl fermions. In a Weyl semimetal, each Weyl node is non-degenerate whereas, in a Dirac semimetal, the Weyl nodes are degenerate as a result of time-reversal and inversion symmetries. Angle-resolved photo-emission spectroscopy has revealed the existence of Dirac nodes in $(Bi,In)_2Se_3$, Na_3Bi and Cd_3As_2, and of Weyl nodes in the TaAs family and in $YbMnBi_2$. The monopoles which are present in topological semimetals can impart interesting transport properties, such as the so-called chiral anomaly, the anomalous Hall effect and the chiral magnetic effect.

There has been much progress in the discovery of Weyl semimetals recently, such as the proposed non-magnetic Weyl semimetal in chalcopyrites including the Ta_3S_2 and $CuTlSe_2$ family of materials, HgTe-class materials, magnetic Weyl semimetals in half-Heusler alloys in a magnetic field and magnetic Heusler alloys, type-II Weyl semimetals in MoP_2 and WP_2.

In type-II Weyl semimetals, Weyl fermions are expected to appear at the boundary of electron and hole pockets. The type-II Weyl semimetals have nearly minimal Weyl nodes and include WTe_2, $MoTe_2$ and their mixtures. In type-II Weyl semimetals, the Weyl fermion appears at the boundary of electron and hole pockets, with a finite density of states near to the Weyl point. This unusual feature of the Fermi surface can lead to a planar orientation-dependent negative magnetoresistance, as in WTe_2 thin films. The negative magnetoresistance of the half-Heusler material, GdPtBi, is instead attributed to band-crossing and to Weyl nodes which appear due to the Zeeman effect. The pressure-dependent negative magnetoresistance of black phosphorus occurs only when the magnetic and electric fields are parallel. It is accompanied by a transition from semiconductor to Dirac semimetal behaviour under hydrostatic pressure.

The Weyl semimetals also exhibit symmetry-protected topological surface states. Bulk Weyl nodes of opposite chirality are the sources and sinks of Berry flux in momentum space, and the projections of these pairs of singularities on a given surface must be linked by open Fermi arcs. The latter have also been detected by means of angle-resolved photo-emission spectroscopy, as have the transport signatures of Weyl semimetal states. Weyl semimetals can resemble Dirac semimetals with regard to transport measurements: just as NbP exhibits one of the highest magnetoresistances (quasi-linear and non-saturating at up to 30T), so too does the Dirac semimetal, Cd_3As_2.

The shot noise and conductance of massless Weyl fermions in a Weyl semimetal resonant junction have been studied[49] by using transfer-matrix methods, showing that those properties can be tuned by varying the barrier strength, the junction structure, the Fermi energy and the crystallographic angle. In a quasi-periodic superlattice, the effect of the degree of disorder strength upon the shot noise and conductance is expected to depend upon a competition between classical and Klein tunneling. The delta barrier structure is also predicted to be critical in determining the shot noise and conductance. A common Fano factor exists in the single delta potential case, while the factor's resonant structure very closely matches the number of barriers in a delta potential superlattice.

One example of accidental quantum-mechanical degeneracy is band-crossing. In the absence of extra symmetry constraints, such an accidental degeneracy is very unlikely to occur in one or two dimensions. In three dimensions however one can occur at isolated points of momentum-space. Also predicted is the existence of a so-called Fermi-arc on a Weyl semimetal surface. This is a unclosed loop which connects projection points of the Weyl node onto the surface Brillouin zone. Time-reversal symmetry-breaking in magnetic Weyl semimetals is unfortunately hard to arrange, because of strong correlation-effects, sample deterioration and other factors. An alternative stratagem is to engineer broken inversion-symmetry when at least four Weyl nodes are present, as in a monocrystal exhibiting no compositional modulations.

Angle-resolved photo-emission spectroscopy is the method-of-choice for directly revealing the energy-band structure and confirming the existence of a Weyl semimetal.

Time-reversal or inversion symmetry-breaking is essential to a Weyl semimetal because there will otherwise occur troublesome double-degeneracies; which explains why a Dirac point can be considered to be the superposition of two Weyl points.

Of particular interest are the magneto-transport properties of Weyl semimetals. Transport in topological semimetals can be grouped into four possibilities, depending upon the strength of the magnetic field. In an essentially zero field, positive magnetoresistance arises from the weak anti-localization effect. In low parallel magnetic fields, a negative magnetoresistance arises due to the non-trivial Berry curvature of topological semimetals. In intermediate-strength fields, a quantum oscillation of the resistivity occurs due to Landau quantization of the energy states. Finally, in a high magnetic field where only the lowest Landau band is occupied, it is unclear whether negative magnetoresistance implies occurrence of the chiral anomaly. In most experiments, a high magnetoresistance is found for a perpendicular magnetic field and sometimes increases linearly with increasing field strength.

The Fermi arcs of Weyl semimetals can be designated as being class-1 or class-2. A tight-binding study of these classes in terms of the tilt strengths of bulk Weyl cones shows[50] that a residual anomalous Hall conductivity of topological surface states is a characteristic of the class-1 arcs. The latter can be deemed to be a non-trivial topological property of hybrid or type-II Weyl semimetals.

A topological semimetal phase having four isolated Weyl nodes in momentum space has been created from a honeycomb arrangement of topological insulator nanowires[51]. In this Weyl semimetal phase, the topological charge response was absent due to the opposed separation of two pairs of Weyl nodes in the Brillouin zone. The topological nature of the system instead manifested itself via a non-zero transverse so-called valley current which was proportional to the length of the Weyl-node separation-vector. An anomalous Hall current could also appear due to the Haldane term; that is, a next-nearest neighbour interwire hopping of electrons in the presence of a modulated magnetic flux.

Weyl and Dirac semimetals are characterized by the chiral anomaly and resultant Fermi-arcs, anomalous Hall effect, negative longitudinal magnetoresistance and planar Hall effect. Study of the occurrence of analogous phenomena in nodal-line semimetals shows that the latter exhibit a three-dimensional analogue of the parity anomaly. This property of two-dimensional Dirac semimetals is manifested at the surface of a three-dimensional topological insulator. The characteristic drum-head property of a nodal-line semimetal has been related to this anomaly, and a field theory was derived[52] which described the correspondingly anomalous response. The planar Hall effect involves the appearance, under conditions in which the usual Hall effect vanishes, of a transverse voltage in the plane of applied non-parallel electric and magnetic fields. On the basis of Boltzmann transport equations, expressions have been derived[53] for the planar Hall effect and longitudinal magnetoresistance in topological insulators in the bulk conduction limit. It was shown that an important role is played by orbital magnetic moments.

A study of the longitudinal and transverse magnetoconductivities of a topological Weyl semimetal near to the Weyl nodes was made by using a two-node model which took account of all of the topological semimetal properties[54]. In a magnetic field, the Fermi energy crossed only the 0th Landau bands of the semimetal phase. Both the longitudinal and transverse magnetoconductivities were positive and linear at the Weyl nodes for a finite potential range, leading to an anisotropic negative magnetoresistivity. The longitudinal magnetoconductivity depended upon the potential range of impurities, and the longitudinal conductivity remained finite in a zero field in spite of the density of states vanishing at the Weyl nodes. It was clear that there was a relationship between the linear magnetoconductivity and the existence of an intrinsic topological Weyl semimetal phase.

As mentioned before, the low-energy quasi-particles of Weyl semimetals are a condensed-matter analogy of the Weyl fermions of relativistic field theory. The non-conservation of chiral charge in parallel electric and magnetic fields – the chiral anomaly – can thus be argued to be the most striking facet of Weyl semimetals, and is attributed to an imbalance in the occupancies of gap-less, zeroth Landau levels of opposite chirality. On the basis of first-principles calculations however, a breakdown of the chiral anomaly in Weyl semimetals has been predicted[55] to occur in the presence of a strong magnetic field. Depending greatly upon the direction of the field, an appreciable energy-gap may open up due to mixing of the zeroth Landau levels associated with separate opposite-chirality Weyl points in the Brillouin zone.

It has been suggested[56] that ideal Weyl semimetal features can be made to coexist with a ferromagnetic ground state in certain compounds having centrosymmetric tetragonal structures. In a magnetic system possessing inversion symmetry, the magnetization direction can be used to manipulate a symmetry-protected band structure into changing from nodal-line type to Weyl type in the presence of spin-orbital coupling. The ferromagnetic nodal-line semimetal is protected by mirror symmetry, with the reflection-invariant plane being perpendicular to the magnetic order. With mirror symmetry breaking being due to magnetization along other directions, the gap-less nodal-line loop will degenerate into just a single pair of Weyl points, protected by rotational symmetry along a magnetic axis which is largely separate in momentum space.

The Weyl semimetals exhibit a dispersion which can be described in terms of a three-dimensional Dirac cone, and which is associated with phenomena such as the chiral anomaly in magnetotransport. The confinement gap in a time-reversal symmetric Weyl semimetal film imparts a mass to the quasiparticles, and time-reversal symmetry is preserved by the existence of an even number of valleys associated with opposite masses. A gate electric field can cause the film to pass through a topological phase transition. The conductivity of a thin film of topological semimetal is determined by interactions of the chiral band structure, the mass term and scalar or spin-orbital scattering. Both of the latter scattering mechanisms exert strong qualitative and quantitative influences upon the conductivity correction. Due to the spin structure of the matrix Green's functions, those terms which are linear in extrinsic spin-orbital scattering are present in Bloch and momentum relaxation times. When the mass term amounts to some 30% of the linear Dirac terms, the system enters the unitary symmetry class of zero quantum correction. Switching the extrinsic spin-orbital scattering then drives the system towards weak anti-localization. In the case of low quasi-particle masses, those terms which are linear in impurity spin-orbital coupling might lead to an observable density dependence of the weak anti-localization correction[57].

The tunnelling conductance of a Weyl semimetal having a tilted energy dispersion was investigated by examining electron transmission through a p-n-p junction comprising one-dimensional electric and magnetic barriers. When both electric and magnetic barriers were present, a large conductance gap could be produced using a tilted energy dispersion without any band-gap. This effect was attributed to a shift in the electron wave-vector at the barrier boundaries. That in turn was caused by a pseudo-magnetic field which was produced by an electrical potential[58]. This is a new feature which occurs only in materials which exhibit tilted energy dispersion.

A study of the zero-temperature alternating-current conductivity of a holographic model Weyl semimetal showed that, at low frequencies, there was a linear frequency-dependence. The model semimetal also underwent a quantum phase transition between a Weyl topological semimetal, exhibiting a non-vanishing anomalous Hall conductivity, and a trivial semimetal. The alternating-current conductivity was characterized by intermediate scaling, due to the existence of a quantum critical region in the system's phase diagram; the latter having been deduced from the scaling properties of the conductivity[59].

Three-dimensional model topological semimetals have been created via the vertical stacking of two-dimensional non-symmorphic lattices, some with and some without breaking crystalline symmetries. Four distinct topological phases could be identified, including Dirac nodal-line semimetals, Weyl nodal-line semimetals, unconventional Weyl semimetals having a topological charge of C = 2, and weak topological insulators. The Weyl nodal loops had no mirror-symmetry protection, and non-trivial drumhead surface states appeared within the loops[60]. When topological phases such as a Weyl semimetal are constructed via the stacking of two-dimensional topological phases, and although the low-dimensional components may be gapped, the higher-dimensional result can be a gapless critical point of a topological phase transition between two different insulating phases. Such a three-dimensional topological semimetal phase has been created by stacking one-dimensional Aubry-Andre-Harper tight-binding lattice models of non-trivial topology. The generalized building-block is a family of one-dimensional tight-binding models having cosine modulations of the hopping and on-site energy terms. A two-parameter generalization of the Aubry-Andre-Harper model can represent topological phases in three dimensions, all within a unified framework. The π-flux state of the two-parameter Aubry-Andre-Harper model involves three-dimensional topological semimetal phases where the topological features are seen in one dimension. This dimensional-reduction technique aids, via the use of simple one-dimensional double-well optical lattices, the experimental construction of the three-dimensional Brillouin zone for a topological semimetal phase[61].

A topological classification scheme, including Chern insulators, has been proposed for generalised Weyl semimetals of any spatial dimensionality and having arbitrary Weyl surfaces which can be non-trivially linked[62]. The scheme modifies already-known three-dimensional constructions in order to account for subtle aspects of semimetal topology. A general charge-cancellation condition for the Weyl surface components was derived by using a fundamental locality principle.

Nodal-Line Semimetals

An early classification of topological semimetals and nodal superconductors which are protected by global symmetries and crystal reflection symmetry depended greatly upon the co-dimension of the Fermi surface or nodal line or point, and whether the mirror symmetry commuted or anti-commuted with non-spatial symmetries. The Fermi surfaces, and nodal lines or points, transformed under mirror reflection and non-spatial symmetries. The classification considered all of the possible symmetry-permitted mass terms that could be added to the relevant Hamiltonian for a given symmetry class, and identified the topological invariants. As compared with point-node Weyl and Dirac semimetals, where the conduction and valence bands touch at discrete points, the two bands cross at closed lines in the Brillouin zone of nodal-line topological semimetals[63]. The two bands cross one another along a closed curve in momentum space. (figure 4).

The curve along which the bands cross is the nodal line. This can be an extended line across the Brillouin zone, or a closed loop inside the zone or even a chain consisting of several connected loops.

Two different types of symmetry-protected nodal line can exist in the absence or presence of spin-orbital coupling. In the case of absence, the nodal line is protected by a combination of inversion symmetry and time-reversal symmetry, but each nodal line has a Z_2 monopole charge and can be created or annihilated only pair-wise. In the presence of spin-orbital coupling, a non-symmorphic symmetry (screw axis) protects a four-band crossing nodal line in systems which possess both inversion and time-reversal symmetry.

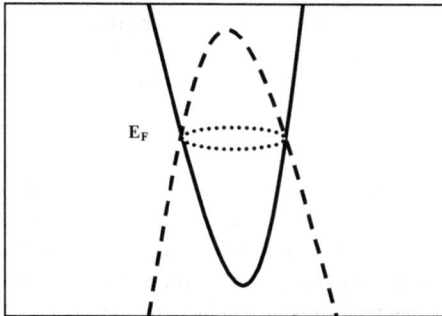

Figure 4. The nodal-line semimetal; the two bands cross one another along a closed curve in momentum space

The crossing of the conduction and valence bands occurs along one or more closed loops in the Brillouin zone, and those loops can be isolated, or may touch at highly symmetrical points. Another type of nodal-line semimetal can exist, which contains a pair of linked nodal loops. A proposed two-band model[64] posited the presence of a pair of nodal lines having a double-helix structure. Because of the periodicity of the Brillouin zone, this could be further twisted into the form of a Hopf link. The nodal lines were here stabilized by combined spatial-inversion and time-reversal symmetries. These two symmetries had to be broken. The band had a non-trivial topology, such that each nodal loop carried a π Berry flux. At the open boundary, surface flat-bands emerged which were completely encircled by the projection of the nodal lines on the surface Brillouin zone.

When the conduction and valence bands cross along a one-dimensional curve in the three-dimensional Brillouin zone, no perturbation which maintains a certain symmetry group via spatial or time-reversal symmetry can remove this crossing line and open up a full direct gap between the two bands. Each nodal line is then topologically protected by the symmetry group, and can be associated with a topological invariant.

The nodal-line topological semimetals involve one-dimensional lines of band-crossing in the Brillouin zone, and the nodal lines can exist in topologically non-trivial configurations. A further twist is the possibility of so-called nodal-knot semimetals in which the nodal lines would form topologically non-trivial knots in the Brillouin zone. Those constructions which produce the simplest trefoil nodal knot, as well as more complicated ones, have been investigated[65]. Yet another possible refinement of the nodal-line semimetal form, is the so-called nodal-link semimetal; involving linked nodal rings in the Brillouin zone. A general construction method, based upon the Hopf map, has been proposed[66].

Recalling that nodal-line semimetals are characterized by the presence of one-dimensional band-touching loops, protected by combined inversion and time-reversal symmetry in the absence of spin-orbital coupling, the loops can also be viewed as being a one-parameter family of Dirac points which exhibit a parity anomaly associated with the above symmetries. This parity anomaly is also manifested by the non-linear optical response in an analogous manner to linear response transport. The presence of a tilting term in the Hamiltonian has been analyzed[67] as an element which does not spoil the above symmetries. It respects the combined symmetry, but breaks each symmetry separately. This offers the possibility of experimentally observing linear and non-linear Hall conductivities in suitable nodal-line semimetals. The linear Hall-like response of tilted nodal-line semimetals is an axion response due to the parity anomaly, extending the class of systems that display such an electromagnetic response.

Yet another form of semimetal, the type-II nodal-line semimetal, has been based[68] upon a two-band cubic lattice model. The zero-energy bulk states here form a closed loop in momentum space, but the local Weyl cones on the nodal line are tilted. Upon considering the repulsive interactions and additional spin degrees-of-freedom, various types of long-range magnetic order appear in the bulk states. An interaction-induced ferromagnetic ordering of the surface states can be expected. At a critical stage in going from a type-I nodal-line semimetal to a type-II nodal-line semimetal, the occurrence of small interactions introduces ferromagnetic order due to the existence of a flat band at the Fermi surface.

A model for a topological semimetal in three dimensions has been proposed[69] in which the energy spectrum exhibits a nodal line that acts like a vortex ring, linked in turn by a pseudo-spin structure which resembles a smoke ring. This vortex ring gives rise to skyrmionic pseudo-spin patterns on both sides of the nodal-ring plane. These patterns cover the entire Brillouin zone and consequently produce a fully extended chiral Fermi arc and a new anomalous Hall effect. Adjustment of the model parameters can shrink the vortex ring until it vanishes, leaving behind a pair of Weyl nodes of opposite chirality. This thus establishes a connection between the momentum-space topologies of a vortex ring (circle of singularity) and of a monopole-antimonopole pair (two point singularities).

A theoretical study[70] has also been made of three-dimensional topological semimetals which have nodal *surfaces* protected by crystalline symmetries, as distinct from nodal points or nodal lines. In these materials, conduction and valence bands cross on closed nodal surfaces in the Brillouin zone. Various classes of nodal surface can be proposed in the absence or presence of spin-orbital coupling. In the former case, a class of nodal surfaces can be protected by space-time inversion symmetry and by sub-lattice symmetry and is characterized by a Z_2 index. Another class of nodal surface is protected by a combination of non-symmorphic two-fold screw-rotational symmetry and time-reversal symmetry. The inclusion of spin-orbital coupling destroys the first class of nodal surface but can preserve the second class if inversion-symmetry is broken. In magnetically ordered systems, protected nodal surfaces can also exist; without and with spin-orbital coupling, provided that certain magnetic group symmetry requirements are satisfied.

A holographic model for a strongly coupled topological nodal-line semimetal predicted[71] that the phase could undergo a quantum phase transition to a topologically trivial state. Use of a dual fermion spectral function showed that there were multiple Fermi surfaces, and that each of them was a closed nodal loop in the semimetal phase. The topological structure of the bulk was produced by an interplay between the dual mass operator and an operator which deformed the Fermi-surface topology. It was concluded that topologically non-trivial semimetal states would generally exist in the presence of strong coupling.

In topological nodal-line semimetals, the band-crossing points form closed loops in momentum space. The materials which fall into this category include very familiar and indeed mundane materials such as face-centered cubic calcium, strontium and ytterbium, hexagonal close-packed beryllium, magnesium, calcium, and strontium, together with compounds such as CaP_3, Ca_3P_2, LaN, $Cu_3(Pd,Zn)N$, $(Tl,Pb)TaSe_2$, $ZrSiS$, $CaAgP$, $CaAgAs$, $BaVS_3$, $BaNbS_3$, $BaTaS_3$, $BaVSe_3$, $BaNbSe_3$ and $BaTaSe_3$. The list also includes more exotic materials, including all-carbon Mackay-Terrones crystals, body-centered orthorhombic C_{16}, Bernal graphite, three-dimensional honeycomb lattices, interpenetrating graphene networks, perovskite iridates and, if stressed, black phosphorus. Their novel properties include almost-flat drumhead-like surface states, and possible 'high-temperature' superconduction. Unusual collective modes result from the peculiarities of the nodal-line structure. Other characteristics include an unique Landau energy level.

Starting with a nodal-line semimetal, and including spin–orbital coupling, a nodal ring can transmogrify into Weyl nodes, into Dirac nodes or, via gap-opening, the semimetal can be converted into a topological insulator.

One class of topological semimetal having point or line nodes can be created when there exists an off-center rotation or mirror symmetry in which the symmetry line or plane, respectively, is displaced from the center of the other symmorphic symmetries of a non-symmorphic crystal. The partial translation which occurs perpendicular to the rotation axis or mirror plane in off-center rotation or mirror symmetry situations then forces two energy bands to stick, and form a doublet in the relevant invariant line or plane of momentum-space[72]. The doublet pair is then a basic feature of any topological semimetal possessing point or line nodes in the presence of strong spin-orbital coupling. The electromagnetic response of gap-less phases in 3+1 dimensions with line nodes involves an intrinsic antisymmetric tensor which is determined by the geometry and energy-embedding of nodal lines[73]. The tensor is simply related to the charge polarization and orbital magnetization; properties which are governed by the geometry of the line nodes.

A family of three-dimensional honeycomb systems has been considered[74] in which the electronic band structures correspond to various topological semimetals having Dirac nodal lines. The latter appeared in various numbers and geometries, depending upon the underlying lattice structure, and were generally stabilized by a combination of time-reversal and inversion symmetries. They were accompanied by topologically protected drum-head surface states. In the bulk, the nodal-line systems exhibited Landau-level quantization and flat bands upon applying a magnetic field. In the presence of spin-orbital coupling, these semimetals became topological insulators.

Methods for identifying the band topology have been based upon monitoring the hybrid Wannier charge centers calculated for Bloch states[75]. They have been applied to crystalline topological insulators and topological semimetal phases, as well as to the rapid screening of materials having non-trivial topologies. On the basis of fundamental connections between topological invariants and the Zak phase, it has been shown[76] that the non-trivial band topologies of two- and three-dimensional topological insulators, as characterized respectively by Chern numbers and Z_2 invariants, are directly reflected by the winding numbers of the Wannier-Stark ladder in an electric field. Floquet Green's function methods indicate that the latter winding number is stable with respect to interband interference and non-magnetic impurity-scattering.

Measurements of Shubnikov-de Haas oscillations suggest that a non-trivial π-valued Berry phase can be expected for Weyl fermions. Such a quantum geometrical phase is normal for all quasi-particles which are associated with a massless linear spectrum. In the presence of strong magnetic fields at very low temperatures, two Weyl nodes of opposite chirality can exchange particles if the Fermi-level lies within the zeroth Landau level. This is the Adler–Bell–Jackiw or chiral anomaly, and is recognised in negative magnetoresistance data when the external magnetic field is collinear with the applied electric field. On the other hand, anomalous negative magnetoresistances have again been observed during measurements of Dirac semimetals such as the above Cd_3As_2, as well as of Na_3Bi and $ZrTe_5$. This makes it unclear whether negative magnetoresistance is due to the intrinsic chiral anomaly or is due to extrinsic factors. It has been proposed that Dirac semimetals can transform into Weyl semimetals due to a lifting of spin degeneracy by the external magnetic field. The fact that Weyl semimetals exhibit the Adler-Bell-Jackiw anomaly implies that the imposition of parallel electric and magnetic fields can pump electrons between nodes of opposite chirality. It can be expected that this pumping is measurable by means of non-local methods in the limit of weak inter-node scattering. Due to the anomaly, application of a local magnetic field parallel to an injected current should introduce a valley imbalance that can diffuse over long distances. A magnetic probe-field could then convert the imbalance into a measurable voltage drop which occurs far from the source and drain. The non-local transport will vanish when the injected current and the magnetic field are orthogonal, thus constituting a test for the chiral anomaly[77]. Similar methods could also be used to characterize Dirac semimetals, where the coexistence of a pair of Weyl nodes at a single point in the Brillouin zone is protected by a crystal symmetry. The existence of an anomaly suggests that valley currents in three-dimensional topological semimetals might be controlled by using electric fields; the latter nodes being analogous to the valley degrees of freedom in semiconductors.

Negative magnetoresistance can also be due to a non-zero Berry curvature, and this possibility subsists for both Dirac and Weyl semimetals. The negative magnetoresistance which is observed is expected to be quadratic in the weak-field limit. In high fields, non-saturating behavior can be exhibited by both types of semimetal. This unexpected field-dependence can persist to above 100K and suggests that current flow in the samples studied is inhomogeneous with regard to the collinearity of the magnetic and electric fields; a geometrical artefact which is termed current-jetting. It can become predominant in negative magnetoresistance measurements of TaAs family members. It has been suggested that, when the fields are parallel, the field-dependent resistivity anisotropy (ratio of the resistivities perpendicular and parallel to the field direction) can explain all of the observed negative magnetoresistance characteristics. It is possible that field-dependent inhomogeneous current distributions can be avoided by using long thin bar-shaped samples in which the electrodes fully cross the sample width. When the defect structures of individual samples have to be considered, analytical modelling of the intrinsic chiral anomaly negative magnetoresistance becomes very difficult. Transmission electron microscopy indicates that those defects are mainly high-density stacking faults in TaAs, and a mixture of stacking faults, vacancies and antisites in TaP.

Energy-band quantum theory has long successfully explained solid-state physical phenomena, especially in semiconductors. During more recent decades, phase-transition and spontaneous symmetry-breaking theories such as Landau-Fermi liquid and Landau-Ginzburg-Wilson have given way to the more exotic Kosterlitz–Thouless, Haldane and quantum-Hall theories. These phase transitions posit non-Fermi liquid behavior, and do not require symmetry-breaking. These new approaches have spurred the rapidly advancing development of topological insulators and topological semimetals. Symmetry-protected phases here arise from the special energy band structure of the material, which obeys Dirac or Weyl theory in a specific momentum space region. The linear energy-band dispersion here exhibits a very different behavior to the expected parabolic energy-band dispersion. The linear energy-band dispersion can moreover be topologically protected, in that the dispersion is preserved provided that a system symmetry is not broken. Topological semimetals do not possess topological order in the usual quantum-topology sense. Like the Haldane phase they instead possess symmetry-protected topological states. The associated transport behavior is also topologically protected, and resists environmental disturbances. The topological semimetals exhibit very different electronic properties as compared with those of metals, conductors and insulators. If combined with a superconductor, for example, the resultant material might find many useful applications.

The integer quantum Hall state was a harbinger of topological matter and eventually led to the topological insulator concept. The integer quantum Hall state's topological nature is best characterised by the Thouless-Kohmoto-Nightingale-den Nijs formula, which links the Berry phase to the Hall conductivity. Topological matter owes its existence to the Berry phase and the realisation that the Berry phase was effectively a new order parameter, which defined topological matter, is regarded as being a major breakthrough in condensed-matter physics.

The topological non-triviality of the topological insulator arises from the existence of a Dirac monopole in a Hamiltonian parameter space, and the concept of topological matter has spread so as to include topological semimetals such as the Weyl and Dirac types. The possible existence of fractional Chern and topological insulators has also been mooted theoretically, due to the analogy with fractional quantum Hall states.

While order parameters are usually related to a spontaneous breaking of the symmetry of the Hamiltonian, the topological order parameter concerns the topological structure of the manifold which is formed by the Hamiltonian parameter. By analogy with superconductors, any closed electron path can be topologically classified in terms of the number of magnetic vortices which it encloses: the number of vortex quanta inside a closed path can be used as a topological order parameter for that path.

As well as angle-resolved photo-emission spectroscopy, transport measurements are an means of investigation of topological semimetals. Giant positive magnetoresistance Shubnikov-de Haas oscillation measurement is commonly used to confirm the presence of unusual phases in materials, such as graphene or a topological insulator, where the energy band exhibits linear energy dispersion. Shubnikov-de Haas oscillation arises from the Landau quantization of electronic states in a high magnetic field; the magnetoresistance oscillates, with a period that depends upon the inverse of the magnetic field strength, as the Fermi-level crosses various Landau levels. The motion of electrons can also lead to a non-zero Berry phase and, particularly when the energy-band exhibits linear dispersion, an extra Berry phase of π can be introduced, leading to other subtleties of interpretation. Thus although Shubnikov-de Haas oscillation measurement is a leading transport-based exploratory technique, it is necessary to treat phase-shift data with caution. The anisotropic behavior of Shubnikov-de Haas oscillations has been used to determine the anisotropic geometry of the Fermi surface of materials such as Cd_3As_2.

As well as Shubnikov-de Haas oscillations, a giant linear magnetoresistance is observed when the imposed magnetic field is perpendicular to the driving current. This phenomenon is attributed to a protection-mechanism which strongly suppresses back-scattering in a zero magnetic field. It results in a high mobility and in a transport lifetime which is 10000 times longer than the quantum lifetime. Removal of that protection by an applied magnetic field then leads to a very large magnetoresistance, possibly due to changes in the Fermi surface which are produced by the applied field.

Two mechanisms have been proposed in order to explain the linear magnetoresistance behavior. In the Parish-Littlewood theory, the linear magnetoresistance is attributed to the occurrence of large mobility fluctuations. In the Abriskosov theory a linear magnetoresistance appears, in a gap-less semiconductor exhibiting a linear dispersion

Materials Research Forum LLC

doi: http://dx.doi.org/10.21741/9781644900154

relationship, when all of the electrons fill the lowest Landau-level. This theory seems untenable when linear magnetoresistance is observed in very low magnetic field.

The most curious property of the bulk states of topological semimetals is the chiral anomaly. When a magnetic field is applied, the Landau levels redistribute. In particular, the zeroth Landau level disperses linearly and the slope corresponds to the chirality of the Weyl node. It can be deduced that the charge in each of the chiral Landau bands is not conserved, and that the magnetic field must be parallel to the electric current in order to produce charge imbalance in the Weyl nodes; thus introducing the chiral anomaly. The carrier density must not be too large however or the $n = 0$ landau level contribution will be smeared out.

The charge imbalance which is introduced between various Weyl nodes by the dot-product of the electric current and magnetic field term requires large momentum scattering to occur in order to relax. If the scattering from one Weyl point to another is negligible or if the inter-node scattering time is sufficiently long, a longitudinal current due to the chiral anomaly can lead to negative magnetoresistance.

In an approximately zero magnetic field, there is a cusp-like conductivity maximum which is attributed to weak anti-localization. The latter occurs in both parallel and perpendicular magnetic fields, and can be explained in terms of a Berry phase of π because the phase difference between two time-reversed routes is the same with the Berry phase circulating along the loop. An accumulation of that phase can suppress back-scattering between two time-reversed routes, and contribute to the conductivity. As well as weak anti-localization, there can be an increase in magnetoconductivity above a certain magnetic field strength in some materials when its direction is parallel to the electric field. This is attributed to chiral anomaly-induced negative magnetoresistance, in that rotation of the magnetic field from parallel to perpendicular with respect to the electric field, first reduces the negative magnetoresistance and then changes it to positive. Quite a low carrier density is required in order to observe the chiral anomaly effect. Conversely, a resistance versus temperature plot can furnish information concerning the carrier density. When the temperature is relatively high, transport is dominated by thermally activated carriers, semiconductor-like behavior is observed because the carrier density is very low and the Fermi level is near to the Dirac point. In the low-temperature region, transport is dominated by the intrinsic carrier density near to the Dirac point and metallic behavior occurs.

Recent research has provided a consistent explanation for the negative magnetoresistance of three-dimensional topological insulators. The negative magnetoresistance which is observed when a magnetic field is applied in the direction of the current is generally attributed to the chiral anomaly, but the latter is not well-defined in the case of topological insulators. The magnetoresistance of a three-dimensional topological insulator was calculated in terms of semi-classical equations of motion; the Berry curvature here introducing an anomalous velocity and orbital moment. The latter, together with g-factors, played an important role. These theoretical results were in quantitative agreement with experimental observations[78]. In particular, the negative magnetoresistance was insensitive to temperature and increased as the Fermi energy approached the band edge.

Many alternatives have to be ruled out in order to confirm that a negative magnetoresistance is due to the chiral anomaly effect. The so-called current jetting effect, a highly non-uniform current distribution, can produce negative magnetoresistance, as in the case of TaAs-type materials. In addition, in the ultra-quantum limit, certain impurities are predicted to be capable of inducing negative magnetoresistance.

As noted elsewhere, the topology of the semimetals makes itself felt not only in the bulk but also in the surface states known as Fermi-arcs; a clear illustration of the non-trivial topological properties of these materials. To reiterate, on their Fermi surfaces there are open arcs, rather than closed loops, which connect the projections of bulk Weyl points on the Brillouin zone surface.

The Fermi-arcs can also be seen as the edge state of Chern insulators; because different Weyl nodes have differing chiralities, the Chern number would be non-zero when only odd Weyl nodes were enclosed and zero when pairs of Weyl nodes were enclosed. A topological phase transition can occur at the boundary of two systems having differing Chern numbers, where the Fermi-arc appears.

Another viewpoint is that the surface dispersions of topological semimetals map onto helicoidal structures: bulk nodal points are projected onto the branch points of the helicoids, and the surface states which are near to various Weyl nodes of opposite chirality correspond to a helicoid or anti-helicoid structure. They constitute an iso-energy contour between different Weyl nodes: the Fermi-arcs. In the case of a Dirac semimetal, a Dirac point is then seen as the superposition of two Weyl nodes of opposite chirality. The surface states near to the projection of each Dirac point therefore the superposition of a helicoid and anti-helicoid which cross along certain lines and can have two associated Fermi arcs. The Fermi arcs can be lost, due to hybridization along the crossing lines, depending upon the symmorphic symmetry.

A path has been proposed, for entropy transport, which involves the Fermi arcs on a given surface being connected to those on the other surface via bulk Weyl monopoles[79]. The low-energy excitations of Weyl topological semimetals generally comprise linearly dispersive Weyl fermions which serve as monopoles of Berry curvature in the momentum space of the bulk material. Topologically protected Fermi arcs on the surface are located at the projections of the Weyl points. The currents which would circulate through the proposed path result in net entropy transport with no net charge transport, and theoretical results have been predicted for Fermi-arc magnetothermal conductivity in classical low-field and quantum high-field limits.

Topological properties are inherently resistant to perturbation and promise the ability to store and process information as state variables. Topological non-trivial states depend upon material parameters such as crystalline symmetry, electron correlations, spin–orbit interactions, and magnetic interactions. Oxide materials offer a wide freedom of choice with regard to such properties. In three-dimensional topological insulators, the conduction- and valence-band characteristics are inverted due to strong spin-orbit interactions. This creates a helical surface or edge state within the bulk gap, and the topological invariant is Z_2 as calculated from the spatial variances of wave functions in momentum space. The surface state is protected against perturbations such as lattice disorder which do not break time-reversal symmetry. Surface electronic states can be directly observed by means of angle-resolved photo-emission and scanning tunnelling spectroscopy. Back-scattering of surface electrons is prohibited because its spin is tightly locked perpendicular to its momentum. A topological insulator which possesses a large bulk gap can permit exploitation of the dissipation-less edge current at above room temperature.

A steep linear dispersion of the electron band indicates the formation of a Dirac cone in the bulk state, recalling a symmetry-protected nodal line. Theory suggests that various topological phase transitions, such as a nodal-line Weyl semimetal, can be created - depending upon the direction of the applied magnetic field - by breaking time-reversal symmetry on the degenerate nodal-line Fermi surface.

The formation of a metallic surface state is only to be anticipated at the hetero-interface between topologically trivial and non-trivial electronic states, but other types of surface state are possible in the case of pyrochlore iridates. Even if the antiferromagnetic domain is not itself in the Weyl topological semimetal phase, and instead in the insulator phase, the surface state should appear at domain walls due to the requirement of continuity of the respective bands across the Fermi level. These states can be classified into various topologies on the basis of rotational symmetry.

It has been suggested that skyrmions may be controllable by means of gating or photo-doping, and skyrmionic oxides which are capable of Fermi-level tuning are suitable for seeking a quantized topological Hall effect. Modulation of the magnetic exchange interactions in such materials would also be desirable.

The Dzyaloshinskii–Moriya interaction is an essential factor in the canting neighboring spins and creating small skyrmions, and arises from inversion symmetry-breaking; something which tends to occur spontaneously at hetero-interfaces. Heterostructures which consist of oxide materials possessing a wide range of magnetic parameters are thus suitable for creating size-controlled skyrmion states. Atomic-scale skyrmion lattices, and nano-scale skyrmions, have also been found in metal multilayers which combined a ferromagnetic metal with a strong spin–orbit coupling metal. By varying the composition of the ferromagnetic metal layer, skyrmion size and density can be adjusted via changes in the magnetic interactions.

The statistical distribution of the energy levels of a topological semimetal has been used to deduce the main parameters which govern so-called universal conductance fluctuations. The said parameters were found to be the number of uncorrelated bands, the level degeneracy and the symmetry parameter. Knowledge of these parameters permits the prediction of the zero-temperature intrinsic universal conductance fluctuations by using Altshuler-Lee-Stone theory. Numerical calculations showed that the predicted conductance fluctuations of quasi one-dimensional topological semimetals of Dirac or Weyl type agreed with theory. On the other hand, a non-universal conductance fluctuation behavior was found in the case of nodal line semimetals, in that the fluctuation amplitude increased with increasing spin-orbit coupling-strength[80]. Unexpected parameter-dependent conductance fluctuations were attributed to the shape of the Fermi-surface.

Having looked at the general theoretical aspects of the subject, it is interesting to look in detail at the various materials which exhibit topological semimetal behaviour.

Antimony and Antimonides

Antimonene

Antimonene is in a different class to that of other two-dimensional crystals[81] due to its strong spin-orbit coupling and the sharp changes in its properties in going from monolayer to few-layer situations[82]. Considering firstly the bulk element, such topological materials host protected surface states with locked spin and momentum degrees of freedom and the helical Dirac nature of the surface states arises from an interplay of the bulk band structure and surface Rashba spin-orbital interaction. This semimetal is a pristine test-bed for investigating the Rashba origins of Dirac-like topological surface states. Momentum-resolved scanning tunnelling spectroscopy over a 300meV energy range has revealed[83] several features that are characteristic of the appearance of Dirac-like surface states from Rashba-type parabolic dispersion. The negative band-gap resulted in a cross-over of the surface-state band structure from Rashba-like at low energies to Dirac-like at higher energies; thus providing a link between conventional Rashba bands and an isolated Dirac cone. One notable feature arising from the dual Rashba-Dirac nature of the surface states was a cross-over behavior in the Landau level and quasi-particle interference dispersions. First-principles calculations demonstrate[84] that, due to a subtle interplay between quantum confinement and surface effects, the topological and electronic properties of Sb(111) nanofilms undergo various transitions as the film thickness is reduced: transforming from a topological semimetal to a topological insulator at 7.8nm (22 bilayers), then to a quantum spin Hall phase at 2.7nm (8 bilayers) and finally to a normal (topologically trivial) semiconductor at 1.0nm (3 bilayers). The electronic structure of a film thus depends upon the number of layers, and can be modified by imposing an epitaxial strain, by changing the effective spin-orbital coupling strength and by choosing the manner in which the layers are stacked. First-principles calculations[85] of double layers again revealed various phases, including topological insulators, topological semimetals, Dirac semimetals, trivial semimetals and trivial insulators. These could be chosen by varying the lattice strain and effective spin-orbital coupling. Spin-orbital coupling in density functional theory band-structure calculations can be treated by using Gaussian basis sets[86]. This involves the evaluation of the radial and angular parts of spin-orbital coupling matrix elements. The results are in good agreement with those of previous methods for all electron basis sets where the full nodal structure was present in the basis elements. In the case of pseudopotential basis sets which lacked a nodal structure, a suitable increase in the effective nuclear potential could encompass all of the relevant band-structure variations which are introduced by spin-orbital coupling. The non-relativistic or scalar-relativistic Kohn-Sham Hamiltonian are first obtained, and the spin-orbital coupling term is added afterwards. As an example, the method was applied to monolayer and multi-layer Sb(111). Most recently it has been shown[87] that an epitaxial interface between one to three atomic layers of lead nano-island, and an Sb(111) substrate, exhibits an extremely high electronic transparency. This causes the lead to lose its superconduction below 1.5K, and also its metallic properties. Its electronic spectrum matches moreover the semimetallic density-of-states of the antimony. Further states arise from a threefold-symmetrical potential, producing 1.8nm periodic modulations of the electronic density, which develops at the lead/antimony interface, and this potential arises from the corrugated structure of the interface.

Topological Semimetals   Materials Research Forum LLC
Materials Research Foundations **48** (2019)   doi: http://dx.doi.org/10.21741/9781644900154

CeSb

A study of electronic transport in magnetic members of the semimetals has shown that extreme magnetoresistance is strongly affected by magnetic order. In particular, the present material exhibits an extreme magnetoresistance of more than 1600000% in a field of 9T. The magnetoresistance itself is non-monotonic across the various magnetic phases and undergoes a transition, from negative to extreme magnetoresistance, as a function of the field when above the magnetic ordering temperature. The magnitude of the extreme magnetoresistance is greater than that for other rare-earth monopnictides, including non-magnetic ones, and obeys a non-saturating power-law in fields exceeding 30T. The overall response has been explained[88] in terms of a modulation of the conductivity by the cerium-orbital state. At intermediate temperatures, it has been explained in terms of an effective-medium model. A comparison with the behaviour of orbitally quenched GdBi supports the assumption of a correlation between extreme magnetoresistance and the onset of magnetic ordering and compensation. The combination of orbital inversion and type-I magnetic ordering in CeSb was deduced to determine its large response.

$CoSb_3$

This skutterudite has a band structure with a topological transition[89] and a symmetry-preserving sub-lattice displacement of antimony atoms very near to the structural ground-state. The cobalt appears in a d^8 configuration with four remaining electrons in the Co_4Sb_{12} unit cell going into other bands. The p_x and p_y orbitals of the Sb_4 ring form either strongly bonding (occupied) or antibonding (unoccupied) states which account for two of the three 5p electrons of each antimony atom. This makes a total of twelve occupied σ-bonding states of each spin. Those twelve π-oriented p_z orbitals couple units as well as cobalt 3d orbitals to form effective molecular orbitals which are centered on the vacant 2a sites of the skutterudite lattice. Eight of those molecular orbitals are occupied by sixteen electrons: one coming from each antimony atom and one coming from each cobalt atom. The topological transition is unusual in that it occurs in a three-dimensional crystal, the band critical point occurs at k = 0, the linear bands are degenerate and the critical point demarcates a trivial phase from a topological phase: essentially a zero-gap topological semimetal. The inclusion of spin-orbit coupling and uniaxial strain changes the system into a topological insulator. The linearity of the band within the gap is the result of a near-degeneracy which also leads to a small band-gap; a characteristic that can make the material a transparent conductor.

InAsSb

It was shown[90] that wires which were made of certain ordered $InAs_{1-x}Sb_x$ compositions exhibited a spin-splitting which was up to twenty times greater than that found in pristine InSb wires. For $x = 0.5$, having a stable ordered CuPt structure, there was an inverted band ordering which led to a novel form of topological semimetal having triple-degeneracy points in the bulk spectrum and topological surface Fermi arcs. Straining could drive the compound into a topological insulator state or could restore normal band ordering. Thus ordered $InAs_{0.5}Sb_{0.5}$ is a semiconductor which exhibited high bulk spin-splitting.

LaSb

This material has a simple rock-salt structure with no broken inversion-symmetry, and exhibits[91] perfect linear band-crossing plus perfect electron-hole symmetry. It nevertheless shares all of the complex field-induced behaviours of more complex semimetals. There is, for instance, a field-induced universal topological-insulator resistivity with a plateau at roughly 15K, together with an ultra-high carrier mobility in the plateau region, quantum oscillations having the angle-dependence of a two-dimensional Fermi surface and an extreme magnetoresistance of about 1000000% in a field of 9T. A phase diagram of the magnetoresistance behaviour can be found under the entry for LaBi.

LuSb

An investigation of the Fermi surface and electrical transport properties showed[92] that, at low temperatures, the magnetoresistance exceeded 3000%, without saturation, in fields of up to 9T (figure 5). Analysis of Hall effect and of Shubnikov-de Haas oscillations showed that the Fermi surface consisted of several pockets which originated from a fairly compensated multi-band electronic structure, in agreement with first-principles calculations. Kohler scaling of the magnetoresistance accounted for the temperature-related behavior. The angle-dependent magnetoresistance could be scaled with the effective mass anisotropy, in perfect agreement with electronic structure and quantum-oscillation analysis. The magnetotransport properties were attributed to the topology of the three-dimensional Fermi surface and to a compensation of electron and hole contributions. The magnetotransport properties could therefore be accounted for without proposing the existence of any topologically non-trivial electronic features. This fitted in with the lack of band inversion and the mixed d−p orbital texture.

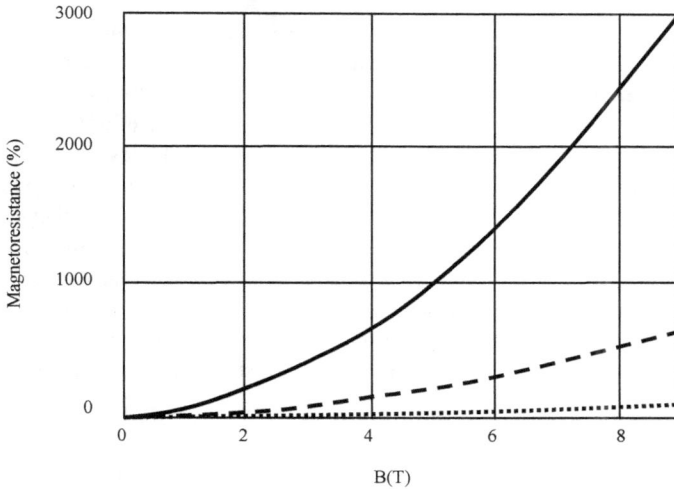

Figure 5. Transverse magnetoresistance of LuSb
Solid line: 2K, dashed line: 50K, dotted line: 100K

Molecular beam epitaxy, core-level and angle-resolved photo-emission spectroscopy, scanning tunnelling microscopy and density functional theory have been used[93] to determine the phase diagram for some surface reconstructions of the (001) plane of CoTiBi: a semiconducting half-Heusler alloy. A simple electron-counting criterion was derived which explains the atomic and electronic structures within a simple framework. This was then applied to the possibly relevant topological semimetal, PtLuSb. In half-Heusler alloys of the form, PtLnQ (Q = Sb or Bi), measurements have confirmed the existence of topological surface states exhibiting novel spin-momentum locking. The (001) surface however exhibits trivial surface states, near to the Fermi energy, which could act as parasitic conduction channels or hybridize with topological states. Upon applying the model, it was expected that dimer stability would occur for half-filled or filled dangling bonds, and was consistent with the c(2 x 2) surface of PtLuSb and the presence of Sb-Sb dimers. Previously observed trivial surface states were attributed to dangling bonds, and could be pushed well below the Fermi energy by controlling the surface lanthanide concentration or surface electron doping.

NbSb$_2$

Magnetotransport properties have been studied[94] in monocrystalline semimetals having a centrosymmetric C12/m1 space group, paramagnetic ground state and non-saturation parabolic-like magnetoresistance. The crystals exhibit metallic conductivity down to 2K, and undergo a metal-to-insulator type of transition in a magnetic field of greater than 4T. They exhibit a resistivity plateau at temperatures below 10K, and the resistivity strongly depends upon the magnitude and direction of the magnetic field. While sweeping the magnetic field from 0 to 14.5T in a transverse configuration at 1.5K, the crystal exhibits a positive magnetoresistance of 4200% in a field of 14.5T together with Shubnikov-de Haas oscillations. A phase diagram of the magnetoresistance behaviour can be found under the entry for LaBi. Hall measurements indicate that carrier compensation between electrons and holes, plus a high mobility and long carrier mean-free-path, contribute to the high magnetoresistance. The Fermi surface of the crystal is quasi two-dimensional, with three-dimensional components.

NdSb

Magnetoresistance and specific-heat measurements showed[95] that this material has a magnetoresistance of about 10000%, even within the antiferromagnetic state. This indicates that extreme magnetoresistance can occur independently of the absence of time-reversal symmetry-breaking in a zero magnetic field. In an applied magnetic field, there is a first-order transition below the Néel temperature. A Dirac semimetal state has been obtained in this material, which has an antiferromagnetic ground state. Its occurrence is supported by band-structure calculations and by the experimental observation of a chiral-anomaly induced negative magnetoresistance[96]. Field-induced Fermi surface reconstruction occurs in response to the change in spin polarization.

SrMnSb$_2$

This is a new type of magnetic semimetal of the form, $Sr_{1-y}Mn_{1-z}Sb_2$ (y,z < 0.1). Strong Shubnikov–de Haas and de Haas–van Alphen oscillations were observed[97], and their analysis demonstrated that the material contains almost massless relativistic fermions together with a non-trivial Berry phase. Neutron scattering measurements revealed a ferromagnetic transition at 565K, followed by a 304K transition into a canted antiferromagnetic state having a net ferromagnetic component.

TaSb$_2$

This topological semimetal has a base-centered monoclinic centrosymmetric structure, undergoes a metal-insulator-like transition in a magnetic field and exhibits a clear resistivity plateau below 13K. Ultra-high carrier mobility and an extremely large magnetoresistance for longitudinal resistivity are observed[98] at low temperatures, together with a quantum oscillation behavior with non-trivial Berry phases. The magnetoresistance is without saturation and the metal-insulator transition occurs whether or not the magnetic field is parallel to the c-axis or the a-axis. The magnetoresistance for B‖c is almost twice as large as that for B‖a at low temperatures. A Kohler-type rule is obeyed by the magnetoresistance at some temperatures when B‖c, but is slightly broken for B‖a. A negative magnetoresistance is observed at up to 9T when the applied field is parallel to the current direction. Angle-dependent magnetoresistance measurements indicate a two-fold rotational symmetry below 70K, reflecting the monoclinic crystal structure. The Hall resistivity exhibits the almost linear field-dependence that implies electron-hole non-compensation behavior. The dumbbell-like angle-dependent magnetoresistance suggests[99] that a strongly anisotropic Fermi surface exists at low temperatures. It loses two-fold symmetry at above 70K, implying a possible topological phase transition. A highly anisotropic transverse magnetoresistance along different crystallographic directions is detected upon rotating the magnetic field[100]. Magnetization measurements revealed strong de Haas-van Alphen oscillations for magnetic fields which were applied along the a- and b-axes, as well as perpendicularly to the ab-plane of crystals. Three Fermi pockets were identified by analyzing the oscillations. Upon applying magnetic fields along various crystal directions, the cross-sectional areas of the Fermi pockets were found to be quite different. Three-band fitting of the electrical and Hall conductivity indicated the presence of two high-mobility electron pockets and one low-mobility hole pocket. *In situ* high-pressure synchrotron X-ray diffraction and electrical transport measurements have been performed[101] using diamond anvil cells and pressures of up to 63GPa. No clear trace of structural phase transition was detected in the diffraction data. The fitting of volume versus pressure data, by using the third-order Birch–Murnaghan equation-of-state, indicated a bulk modulus of 131.2GPa, and a first-order derivative of 6. Between ambient and 27.8GPa, the low-temperature conduction behavior was essentially unchanged, and could be described by a power law having an exponent of about 3. The results implied that the topological-semimetal state was stable against pressures of up to at least 27.8GPa.

YSb

This material exhibits a magnetoresistance of more than 1000% and a low-temperature resistivity plateau, thus making it a suspected topological semimetal. Analysis of its Fermi surface by means of band calculations, magnetoresistance and Shubnikov-de Haas effect measurements revealed only three-dimensional Fermi sheets. The application of

Kohler scaling to magnetoresistance data accounted for its low-temperature behavior. Angle-dependent magnetoresistance measurements[102] indicated a three-dimensional scaling which produced an effective-mass anisotropy that agreed perfectly with the electronic structure and quantum-oscillation analysis, thus providing support for a three-dimensional Fermi-surface explanation for magnetotransport and casting doubt on the need to propose the existence of topologically non-trivial two-dimensional states.

Arsenides

A topological phase transition has been predicted to take place in fields of about 4T in high-mobility GaAs artificial two-dimensional lattices of triangular symmetry, and to be detectable due to anomalous features of the edge conductance[103]. A quantum phase transition between topological insulating and topological semimetal states can be driven by an in-plane magnetic field. The semimetal has band-touching points conveying a quantized Berry flux, and edge-states which terminate at the touching points.

In transition-metal arsenides, the conduction and valence bands touch to give a gap-less bulk Dirac dispersion at the pair of Dirac points which are located along a rotational axis and are protected by the associated rotational symmetry. As noted elsewhere, a helical surface state (Fermi-arc) appears to link the Dirac points. Overall, the Dirac semimetal encompasses two- or three-dimensional topological insulators, the Weyl topological semimetal and topological superconductors, by controlling dimensionality and symmetry, etc. Upon breaking the time-reversal or space-inversion symmetry, the Dirac points split into two Weyl points and a Weyl topological semimetal appears. External fields can control the topological phase domains and the related dissipation-less edge states.

CaAgAs

The magnetotransport properties of the purported nodal-line semimetals, CaAgAs and CaCdGe, have been studied in monocrystalline and polycrystalline samples[104]. The results could be explained in terms of the single-band model, and were consistent with the prediction that only non-trivial Fermi pockets - linked by the topological nodal-line - are present at the Fermi level. First-principles calculations indicated that CaAgAs is a single-band material having one donut-like hole Fermi pocket. This was consistent with its being the so-called first hydrogen-atom nodal-line semimetal. Similar calculations showed that CaCdGe has one donut-like hole Fermi pocket, which originates from the band exhibiting the nodal-line feature, and one trivial ovoid-like electron Fermi pocket. The magnetotransport properties are therefore quite different. In fields of 2 and 9T, a linear transverse magnetoresistance of up to 18% was observed in CaAgAs, whereas an extremely large non-saturating quadratic magnetoresistance of up to 3200% was found for CaCdGe. This suggested that the electron-hole compensation effect was responsible for the extremely large values observed in CaCdGe. Angle-dependent Shubnikov-de Haas oscillation data for CaCdGe indicated the presence of a Fermi pocket with an oscillation frequency of 204T and effective mass of $0.23m_e$; thus agreeing well with the ovoid-like electron Fermi pocket prediction. The use of angle-resolved photo-emission spectroscopy and bulk-sensitive soft X-ray techniques has experimentally demonstrated[105] that the hexagonal pnictide, CaAgAs, is a topological insulator which is characterized by its possession of an inverted band structure and mirror reflection symmetry. There is a bulk valence-band structure in the three-dimensional Brillouin zone, and a Dirac-like energy band and ring-torus Fermi surface are associated with a line node where the bulk valence and conducting bands cross in momentum space in the presence of negligible spin-orbit coupling. No other bands cross the Fermi level and the low-energy excitations are thus associated only with the Dirac-like band.

Cd_3As_2

In n-doped Cd_3As_2, a non-saturating linear magnetoresistance was observed in a magnetic field of about 65T. This was attributed to disorder effects rather than to the above protective mechanism. The latter magnetic field drove Cd_3As_2 to the quantum limit, with no discernible change in the Fermi surface. In the case of the Dirac semimetal, TlBiSSe, a large linear magnetoresistance was attributed to the Hall field. A large non-saturating magnetoresistance has been observed in NbP, but this material has a band structure which is different to that of the usual Weyl semimetal in that the band structure comprises hole pockets arising from normal quadratic bands as well as electron pockets arising from linear Weyl bands. Its large magnetoresistance is therefore attributed to electron–hole resonance. When the applied magnetic field exceeds a critical value, the time-reversal symmetry is broken and a Dirac semimetal is turned into a Weyl semimetal.

The energy-band structure, assuming a hypothetical fluorite structure was first determined[106] by using the pseudopotential method, leading to a qualitative picture of the optical and transport properties. The crystal structure was known to be body-centered tetragonal, with eighty atoms per unit cell. This structure can be understood by looking along the c-axis (figure 6). This shows that each cadmium is tetrahedrally coordinated with arsenic-atom nearest-neighbors, while each arsenic atom is surrounded by cadmium atoms which are located at six of the eight corners of a cube; two vacant sites being located at diagonally opposite corners of a cubic face. The space group is such that some of its symmetry operations are associated with non-primitive translations. The first Brillouin zone was also deduced (figure 7).

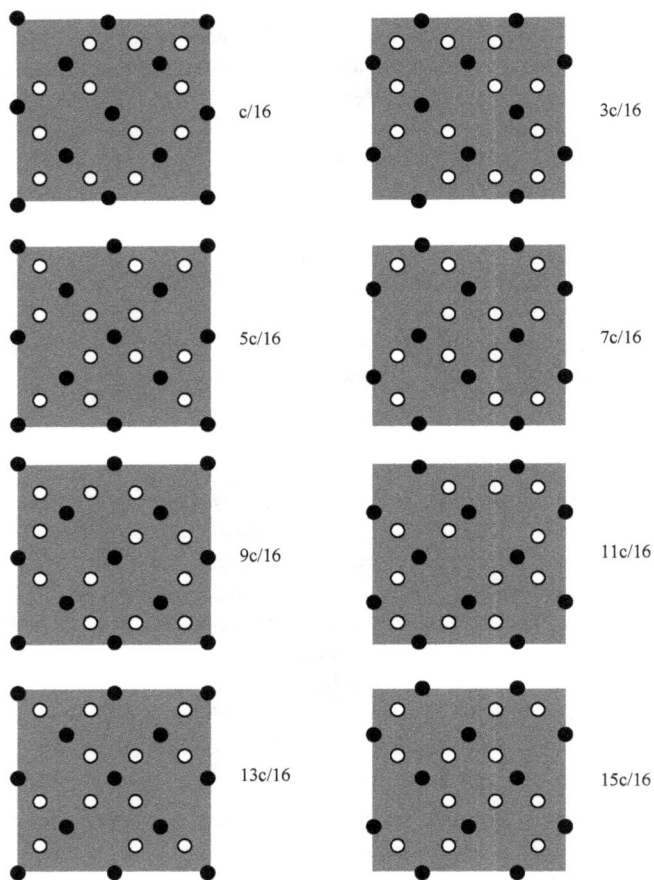

Figure 6. Crystal structure of Cd₃As₂, viewed on 8 equal planes along the c-axis
Cadmium atoms are represented by open circles on each plane, while
arsenic atoms which are 1/16 below the plane are represented by solid circles

The structure is complicated and, although it can be related to a tetragonally distorted antifluorite structure with ¼ cadmium site vacancies, at temperatures greater than 873K, the distribution of those vacancies is random and leads to the ideal antifluorite space-group, Fm$\bar{3}$m. Between 873 and 648K, the cadmium ions are ordered, leading to the structure, P4$_2$/nmc. At lower temperatures, the crystal structure was initially thought to be non-centrosymmetric, I4$_1$cd, but was finally declared to be centrosymmetric, I4$_1$acd. This ambiguity between centrosymmetric and non-centrosymmetric structures is difficult to resolve in some cases. Factor-group analysis indicates 145 Raman-active phonon modes for non-centrosymmetric I4$_1$cd and 74 modes for centrosymmetric I4$_1$acd.

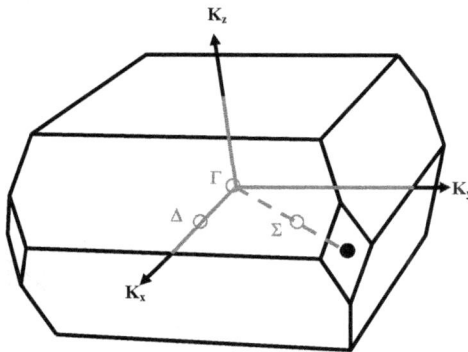

Figure 7. The first Brillouin zone of the body-centered tetragonal Cd$_3$As$_2$ structure

Other early work on this material[107] yielded electron effective-mass values (figure 8) via low-temperature Shubnikov-de Haas, magneto-Seebeck and Hall measurements. Using Kane's model for a HgTe-type inverted energy-band structure, the dispersion relationship for the conduction band was obtained together with a Γ_8-Γ_6 energy-gap of 0.11eV. The heavy-hole valence band was separated from the conduction band at Γ by a residual gap of about 0.01eV. There was a second conduction band, minimum of which at Γ was about 0.6eV above the Γ_8 valence band. There was suspected to be a third one which was some 0.4eV higher.

This well-known semiconductor of high carrier mobility has more recently been found to be a symmetry-protected topological semimetal having a single pair of three-dimensional Dirac points in the bulk, and non-trivial Fermi arcs on the surface. It can be made into a topological insulator and Weyl semimetal state by symmetry-breaking. It can also be made into a quantum spin Hall insulator, with a gap of more than 100meV, by reducing its dimensionality[108]. It was expected that three-dimensional Dirac cones in the bulk could support an appreciable linear quantum magnetoresistance; even at up to room temperature.

Topological Semimetals   Materials Research Forum LLC
Materials Research Foundations **48** (2019)   doi: http://dx.doi.org/10.21741/9781644900154

The electronic structure was investigated by means of angle-resolved photo-emission measurements, performed on the (112) surface, and by means of band-structure calculations. The measured Fermi surface and band structure results were in good agreement with the assumption that two bulk Dirac-like bands approached the Fermi level and formed Dirac points near to the Brillouin-zone center[109]. A topological surface state with a linear dispersion approaching the Fermi level was identified. High-resolution angle-resolved photo-emission spectroscopy of (001)-cleaved surfaces revealed that a highly linear bulk band crossing formed a three-dimensional dispersive Dirac cone projected at the Brillouin zone centre[110]. An unusually high in-plane Fermi velocity of up to 1.5×10^6m/s was observed, with a mobility of up to 40000cm^2/Vs, confirming that this material was promising as an anisotropic-hypercone three-dimensional high spin-orbital analogue of three-dimensional graphene.

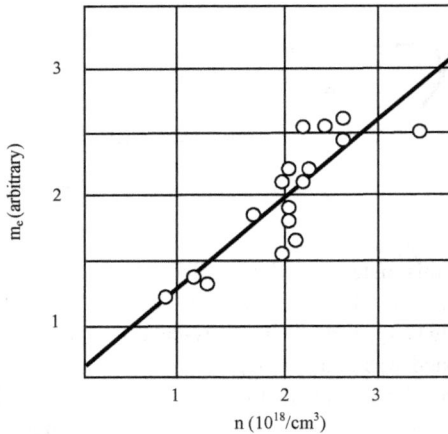

Figure 8. Variation of the low-temperature effective mass
in Cd$_3$As$_2$ as a function of the electron concentration

Crystals which were grown along the [100] and [112] directions by means of chemical vapor transfer permitted examination of the transport anisotropy. The resistivity and magnetoresistance were essentially linear with regard to the magnetic field, between 2 and 300K, regardless of direction. The linear resistivity and magnetoresistance were markedly anisotropic with regard to the [100] and [112] directions and also with regard to

Topological Semimetals   Materials Research Forum LLC
Materials Research Foundations **48** (2019)   doi: http://dx.doi.org/10.21741/9781644900154

the tilt angle between the growth directions and the magnetic field[111]. This yielded the transport signatures of the three-dimensional Dirac fermion and the possible linear and anisotropic changes in the Weyl Fermi surface in the magnetic field. A very large magnetoresistance along the [100] direction approached 3100% at 2K in 14kOe. The Hall anomaly demonstrated two-carrier transport which was accompanied by a transition from n-type to p-type conduction with decreasing temperature. The carrier-type transition was explained by considering temperature-dependent spin-orbital coupling. The magnetoresistance had a non-saturating value of up to 2000% at high temperatures and was attributed[112] to electron-hole compensation. The present Dirac semimetal has already held the record for high mobility and for positive linear magnetoresistance in perpendicular magnetic fields. A large negative magnetoresistance is expected in topological semimetals, when placed in a parallel magnetic field, reflecting a chiral anomaly.

Table 1. Sheet and bulk carrier densities and Hall mobility of Cd$_3$As$_2$
thin films as a function of growth conditions, thickness and temperature

Growth	t (nm)	T (K)	n_S (/cm^2)	n_B (/cm^3)	Mobility (cm^2/Vs)
150C, 5min	85	2	1.46×10^{12}	1.7×10^{17}	6700
150C, 5min	85	300	4.25×10^{12}	5.0×10^{17}	4550
150C, 10min	120	2	2.17×10^{12}	1.8×10^{17}	7900
150C, 10min	120	300	7.42×10^{12}	6.2×10^{17}	6400
150C, 20min	170	2	1.43×10^{12}	0.8×10^{17}	5250
150C, 20min	170	300	13.6×10^{12}	8.0×10^{17}	8950
150C, 40min	340	2	3.76×10^{12}	1.1×10^{17}	11250
150C, 40min	340	300	24.1×10^{12}	7.1×10^{17}	11950
170C, 60min	370	2	13.0×10^{12}	3.5×10^{17}	13650
170C, 60min	370	300	52.5×10^{12}	14.2×10^{17}	12050

Observations were subsequently made of negative magnetoresistance in micro-ribbons when placed in parallel magnetic fields at 50K. The negative magnetoresistance was sensitive to the angle between the magnetic and electrical fields, was little affected by temperature and depended upon the carrier density. A large negative magnetoresistance resulted from low carrier densities in the samples; which ranged from 3.0 x 10^{17}/cm^3 at 300K to 2.2 x 10^{16}/cm^3 below 50K[113]. The observed negative magnetoresistance was therefore attributed to the aforementioned chiral anomaly. Although a large non-saturating linear transverse magnetoresistance and a negative longitudinal magnetoresistance are often considered to be evidence of fermions that have a chirality, classical mechanisms can however - due to disorder or non-uniform current injection - also produce a negative longitudinal magnetoresistance. Magnetotransport measurements of epitaxial thin films revealed a quasi-linear positive transverse magnetoresistance and a negative longitudinal magnetoresistance. By using films having various thicknesses (table 1), and by relating the temperature dependence of the carrier density and mobility to the magnetoresistance characteristics, it was demonstrated that the quasi-linear positive magnetoresistance and negative magnetoresistance were due to conductivity fluctuations and that no chiral anomaly need be involved[114]. On the other hand, field-modulated chiral charge pumping and valley diffusion studies[115] revealed, apart from the usual negative magnetoresistance, a non-local response which involved a negative field dependence up to room temperature and which originated from the diffusion of valley polarization. A large magneto-optical Kerr effect was generated by parallel electric and magnetic fields. These new experiments furnished quantitative evidence of the chiral anomaly which was previously unavailable.

In perpendicular magnetic fields, positive linear magnetoresistances up to 1670% in 14T at 2K were observed. When the temperature- and doping-dependent thermo-electric behaviors were calculated using Boltzmann transport theory, the calculated properties of pristine material closely matched experimental results[116]. Hole-doping was especially likely to improve greatly the thermoelectric behavior. The optimum figure-of-merit following electron doping (10^{20}/cm^3) was about 0.5 at 700K. This was much larger than the maximum experimental value found for pristine material. In p-type samples, the maximum value of the Seebeck coefficient as a function of temperature seemed to increase with increasing hole-doping concentration. Its position shifted markedly towards lower temperatures, as compared with that of n-type samples. This led to an optimum figure-of-merit of about 0.5 at 500K in p-type (10^{20}/cm^3). Due to coupling between open Fermi arcs on opposite surfaces, these topological Dirac semimetals exhibit a new type of cyclotron orbit, known as a Weyl orbit, in the surface states. Upon decreasing the carrier density in nanoplates, a cross-over has been observed from multiple-frequency to single-frequency Shubnikov-de Haas oscillations in an out-of-plane magnetic field; thus reflecting the predominant role played by surface transport[117]. With increasing magnetic field, the Shubnikov-de Haas oscillations develop into a quantum Hall state with a non-vanishing longitudinal resistance. By tracking the oscillation frequency and Hall plateau, a Zeeman-related splitting was observed and the Landau level index could be deduced, as well as the sub-band number. Unlike the case for normal two-dimensional systems, this unusual quantum Hall effect was expected to be related to the quantized version of Weyl orbits. Another study[118] of the Shubnikov-de Haas oscillation of Weyl and Dirac semimetals took account of their topological nature and inter-Landau band scattering. In the case of a Weyl semimetal with broken time-reversal symmetry, the phase-shift changes non-monotonically and goes beyond known values of $\pm 1/8$ and $\pm 5/8$ as a function of the Fermi energy. In the case of a Dirac semimetal or paramagnetic Weyl semimetal, time-reversal symmetry leads to a discrete phase-shift of $\pm 1/8$ or $\pm 5/8$. The topological band inversion can lead to beating patterns in the absence of Zeeman splitting. The results were suggested to explain experimental data on the present material.

Raman scattering in Cd_3As_2 involves a complicated interplay of electronic and phonon degrees-of-freedom. Resonant phonon scattering due to interband transitions, anomalous anharmonicity of the phonon frequency and intensity plus quasi-elastic electronic scattering have been observed[119]. The latter two properties are controlled by a characteristic temperature scale of 100K which is related to mutual fluctuations of the lattice and electronic degrees of freedom. The characteristic temperature corresponds to the energy of optical phonons which couple to interband transitions in the Dirac states. Electron-phonon coupling in topological semimetals is related mainly to phonons possessing finite momenta. Back-action on the optical phonons is observed only as anharmonicities via multi-phonon processes which involve a broad range of momenta. First-principles density functional theory studies of the band structure and dielectric function of a body-centered Cd_3As_2 crystal confirmed[120] it to be a Dirac semimetal having two Dirac nodes near to the Γ-point on the tetragonal axis. The bands near to the Fermi level exhibit linear behavior, the resultant Dirac cones are anisotropic and electron-hole symmetry along the tetragonal axis is destroyed. Along that axis, symmetry-protected band linearity exists only within a small energy interval. Behavior which resembles that of a three-dimensional graphene-like material is attributed to arsenic p-orbitals, pointing to cadmium vacancies in directions which vary throughout the unit cell. As a result of the Dirac nodes, the imaginary part of the dielectric functions exhibits a plateau for vanishing frequencies where the finite value is proportional to the Sommerfeld fine-structure constant but depends upon the polarization of the light. An energy gap opens up in the bulk electronic states of sufficiently thin films and carriers residing in surface states dominate electrical transport at low temperatures. The carriers in such states are sufficiently mobile to generate a quantized Hall effect. Sharp quantization reveals a surface transport that is almost free of parasitic bulk conduction[121]. Doping with europium causes a change in the magnetoresistance sign from positive to negative. Measurements of electron spin resonance and magnetic susceptibility show that there are two types of Eu^{2+} magnetic ions which occupy the positions of cadmium ions and tetrahedral vacancies so as to form ferromagnetic and antiferromagnetic phases, respectively[122]. This suggests the occurrence of small-scale phase separation and of transformation of the Dirac semimetal into a Weyl semimetal due to magnetic impurities.

The highly conductive bulk state usually overshadows electronic transport arising from the surface state in a Dirac semimetal. The supercurrent which is carried by bulk and surface states in short and long $Nb|Cd_3As_2$-nanowire$|Nb$ junctions, respectively, was therefore explored[123]. In an approximately 1μm-long junction, Fabry-Pérot interferences-induced oscillations of the critical supercurrent were observed. This suggested ballistic transport of the surface states supercurrent, where the bulk states were decoherent and the topologically protected surface states remained coherent. A superconducting dome was observed in long junctions and was attributed to an enhanced de-phasing arising from the interaction between surface and bulk states as a tuning gate voltage to increase the carrier density. Films have been grown on $SrTiO_3$ substrates, by means of solid-phase epitaxy at up to 600C, using optimised capping layers and substrates[124]. The arsenic triangular lattice was found to be stacked epitaxially on the square titanium lattice of the (001) $SrTiO_3$ substrate. This produced (112)-oriented Cd_3As_2 films of high crystallinity, with a rocking-curve width of 0.02° and an electron mobility greater than $30000cm^2/Vs$. A two-step crystallisation process was identified, in which out-of-plane and then in-plane, directions occurred with increasing annealing temperature. This led to an unique approach to the fabrication of high-quality Cd_3As_2 films and to the investigation of quantum transport by back-gating via the $SrTiO_3$ substrate. As noted, Dirac semimetals exhibit Fermi-arc surface states; represented by a set of discrete surface sub-bands in nanowires, due to quantum-confinement effects. A tunable Fano effect has been induced by interference between the discrete surface states and continuous bulk states of Cd_3As_2 nanowire[125]. The discrete surface bands led to a zero bias peak in conductance as the Femi level was tuned across the surface sub-bands. The Fano resonance resulted in the appearance of an asymmetrical line-shape in the differential conductance spectrum. The Fano interference could introduce an additional phase into the Weyl orbits and lead to modification of the oscillation frequency.

A room-temperature optical reflectivity study, under external pressures of up to 10GPa, has been made[126] of [112]-oriented monocrystals over a broad energy range. A sudden drop in the band dispersion parameter, and an interruption of the gradual red-shift of the band-gap at about 4GPa confirmed the occurrence of a structural phase transition from tetragonal to monoclinic. A pressure-induced increase in the overall optical conductivity at low energies, and a continuous red-shift of the high-energy bands, indicated that the system evolved towards a topologically trivial metallic state. On the other hand, complete closure of the band-gap was not observed within the present pressure range. Investigation of the low-pressure regime implied the existence of an intermediate state, between 2 and 4GPa, that presaged the structural phase transition or was due to a lifted degeneracy of the Dirac nodes. Some of the optical data indicated another anomaly, at 8GPa, which might be associated with low-temperature superconductivity.

In a Dirac semimetal which arises from band inversion, such as the present material or Na_3Bi, the expected double Fermi arcs on the surface are not topologically protected. In general, the arcs deform into states which are similar to those on the surface of a strong topological insulator. The fact that some perturbations are unable to destroy the double Fermi arcs has been attributed to a non-generic extra particle-hole symmetry. In other cases, surface states can be completely destroyed without breaking symmetry or affecting the bulk Dirac nodes. The conclusion[127] is that there can exist bulk Dirac semimetals which have no surface states, although no such example is currently known.

CuMnAs

Monocrystals of orthorhombic CuMnAs, which is predicted to be a topological Dirac semimetal with a magnetic ground state and broken inversion symmetry, exhibit an antiferromagnetic transition at about 312K. The material has a TiNiSi-type structure belonging to the non-symmorphic orthorhombic space group, Pnma. The copper and arsenic atoms form a three-dimensional four-fold connected anionic network which is composed of strongly corrugated two-dimensional sheets of edge-sharing six-membered rings. The manganese atoms fill the large channels which exist along the b-axis. The spin orientations lie along the so-called easy c-axis and the magnetic moments in one channel are aligned antiparallel to each other. Between any two channels, the magnetic moments on the inversion-related manganese atoms are also aligned along the opposite directions. This means that the P- and T-symmetries are broken while the combined PT-symmetry is maintained. The magnetic properties further suggest that the antiferromagnetic order may be tilted with respect to the spin orientation in the bc-plane. A small isotropic magnetoresistance, and a linearly field-dependent Hall resistivity having a positive slope, indicate that single hole-type carriers at a high density but with low mobility dominate the transport properties (figure 9). Low-temperature heat capacity data show that the effective mass of the carriers is much greater than that of those in typical topological semimetals; implying that the carriers in the present material are very different to the predicted Dirac fermions[128].

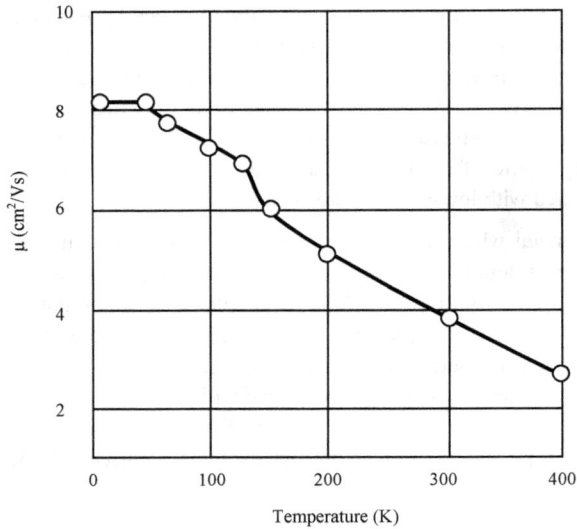

Figure 9. Carrier mobility in CuMnAs as a function of temperature

Further work indicated that the material would be an antiferromagnetic Dirac semimetal if the R_y gliding, and S_{2z} rotational, symmetries were both maintained in the magnetic ordered state. In the low-temperature commensurate antiferromagnetic state, the b-axis is the easy magnetic axis; breaking S_{2z} symmetry and resulting in a polarized surface state. Study of the anisotropic magnetic properties and magnetoresistance of $Cu_{0.95}MnAs$ and $Cu_{0.98}Mn_{0.96}As$ study showed that, in $Cu_{0.95}MnAs$, the magnetic easy axis lay along the b-direction while the hard axis lay along c. This composition also underwent a spin-flop phase transition (figure 10) at high temperatures and low fields when the field was applied along the easy b-axis. No metamagnetic transition was observed in $Cu_{0.98}Mn_{0.96}As$; indicating that the magnetic interactions in the system are very sensitive to the presence of copper vacancies and Cu/Mn site-mixing[129].

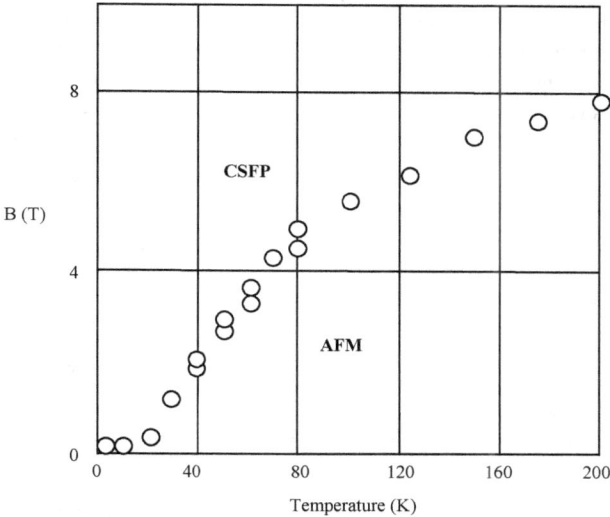

*Figure 10. Magnetic phase diagram of $Cu_{0.95}MnAs$ with the field parallel to the b-axis
CSFP: canted spin-flop phase, AFM: antiferromagnetic*

HfIrAs

A study was made[130] of the lattice constants and topological phases of half-Heusler compounds of the form, HfIrX, where X is arsenic, antimony or bismuth. In the absence of spin-orbital coupling, unstressed cubic crystals of HfIrAs and HfIrBi are non-trivial topological semimetals, while HfIrSb is a trivial topological insulator. This is because the so-called internal pressure lifts the s-type Γ_5 band in HfIrSb. The band-gap between s-type Γ_1 and p-type Γ_5 is very sensitive to the magnitude of the lattice constant. When the lattice constant of HfIrAs is smaller than 5.97Å, it has a topologically trivial band structure with the Γ_1 band above the Γ_5 band. It has a topologically non-trivial band structure when the lattice constant is greater than 5.97Å. When spin-orbital coupling is incorporated, HfIrAs and HfIrSb become a non-trivial topological insulator and an ordinary band insulator, respectively, but HfIrBi remains a topological semimetal. The spin-orbital coupling strength is greater in HfIrAs than in HfIrBi. When a compressive stress is applied to the ab-plane of HfIrBi, there appear eight Weyl points, with four so-called source Weyl points in the K_xK_z-plane at ($\pm0.023/Å$, 0, $\pm0.108/Å$) and four so-called drain Weyl points in the K_yK_z-plane at $(0, \pm0.023/\text{Å}, \pm0.108/\text{Å})$. Density functional theory calculations have recently predicted[131] the possible existence of a new family of topological semimetals of the form, X_2YZ, where X is copper, rhodium, palladium, silver, gold or mercury, Y is lithium, sodium, scandium, zinc, yttrium, zirconium, hafnium, lanthanum, praseodymium, promethium, samarium, terbium, dysprosium, holmium or thulium and Z is magnesium, aluminium, zinc, gallium, yttrium, silver, cadmium, indium, tin, tantalum or samarium. These alloys evidence the existence of multiple triple-point topological fermions, on four independent axes; quasi-particles which have no analogous counterpart elementary particle in the standard model. Simulated angle-resolved photo-emission spectroscopy reveals the complicated topological surface states and characteristic Fermi-arcs. The inclusion of spin-orbital coupling splits the triple-point into two Dirac points. The triple-point fermions are found on the eminently cleavable (111)-surface and well-separated from the surface point, thus permitting them to be resolved by surface spectroscopy.

Hg_3As_2

Two-dimensional topological semimetals are little investigated, and it has been proposed[132] that such a state could be achieved in a mixed lattice comprising Kagome and Honeycomb lattices. It is known that compounds such as A_3B_2, where A is a group-IIB cation and B is a group-VA anion, which possess such a lattice are two-dimensional nodal-line semimetals due to band inversion between the cation mercury s-orbital and the anion arsenic p^2-orbital with respect to the mirror symmetry. Because the band inversion occurs between two bands having the same parity, these peculiar two-dimensional nodal-line semimetals could be used as transparent conductors. In the presence of buckling or spin-orbital coupling, the two-dimensional nodal-line semimetal state can turn into a two-dimensional Dirac semimetal state or a two-dimensional topological crystalline insulating state. Because the band-gap opening due to buckling or spin-orbital coupling is small, the Kagome-Honeycomb lattice can still be regarded as being a two-dimensional nodal-line semimetal at room temperature. Band-inversion in general produces a wide variety of physical effects in both topological semimetals and topological insulators[133]. For example, the inverted band structure which exhibits so-called Mexican-hat dispersion can increase interband correlation and lead to strong intrinsic plasmon excitation in which the frequency ranges from a few, to tens, of meV and can be modified by an external field. The asymmetrical electron-hole term also splits the peak of the plasmon excitation in two.

LaAs

The effects of hydrostatic pressure upon the electronic properties have been predicted[134] using density functional theory calculations and a screened hybrid functional. Particular attention was paid to the band-crossing near to the X-point, which can make this material a topological semimetal. One approach predicted that it would exhibit a crossing, between the highest arsenic p-band and the lowest lanthanum d-band, near to the X-point given the calculated equilibrium lattice parameter. Such a crossing was not predicted when the band overlap was calculated using a different approach. The p-d crossing was predicted to occur regardless when the material is subjected to a hydrostatic pressure, and to exhibit a topological phase transition at about 7GPa. In agreement with experimental observations, the material's rock-salt crystal structure was predicted to be stable under applied pressures of up to 20GPa.

NbAs

In the original experimental discovery of a Weyl semimetal state in this inversion-symmetry breaking monocrystalline solid, a combination of soft X-ray and ultra-violet photo-emission spectroscopy revealed[135] Weyl cones in the bulk and Fermi-arcs on the surface. First-principles calculations of the non-magnetic materials family, TaAs, TaP, NbAs NbP, revealed[136] 12 pairs of Weyl points within the entire Brillouin zone of each of them. In the absence of spin-orbit coupling, band-inversions in mirror-invariant planes led to gap-less nodal rings in the energy-momentum dispersion. The strong spin-orbit coupling then opened-up full gaps in the mirror planes to generate non-zero mirror Chern numbers and Weyl points located off the mirror planes. The resultant surface-state Fermi-arc structures on both (001) and (100) surfaces were noted to have interesting shapes. Systematic electronic band structure calculations revealed[137] the nature of the Fermi surfaces and their complicated interconnectivity in this family (TaAs, TaP, NbAs, NbP) of materials. Similar first-principles calculations of the surface band-structures of those materials revealed[138] Fermi-arcs having a spin-momentum locked spin texture. In the case of the (001) polar surface, the shape of the Fermi surface depended sensitively upon the surface termination (cation or anion) although the arc-topology was the same. Phosphorus- or arsenic-terminated surfaces were consistent with photo-emission measurements, and the surface potential dependence suggested that the shape of the Fermi surface can be manipulated by depositing species such as potassium. On polar surfaces of a Weyl semimetal without inversion-symmetry, Rashba-type spin polarization in the surface states leads to a marked spin texture. By studying the spin polarization of the Fermi surface, it was thus possible to distinguish Fermi arcs from trivial Fermi circles. Given that the four members (NbP, NbAs, TaP, TaAs) of this family exhibit an increasing degree of spin-orbit coupling in the band structure, comparison of their surface states revealed the changes in topological Fermi arcs in going from a spin-degenerate Fermi circle to a spin-split arc as the spin-orbit coupling increases from zero to a finite value.

It has been shown that negative magnetoresistance in the non-centrosymmetric materials, NbP and NbAs, can be caused by the intrinsic chiral anomaly of Weyl fermions or by effects such as the superposition of Hall signals, field-dependent inhomogeneous current flow and the weak localization of coexistent trivial carriers. This diversity can be problematic, given that chiral-anomaly induced negative magnetoresistance is widely used as a key criterion for identifying the existence of Weyl fermions in topological semimetals. The existence of weak-localization controlled negative magnetoresistance is very dependent upon the sample quality and is revealed by a marked cross-over from positive to negative magnetoresistance changes at high temperatures. This results from a competition between the phase coherence time and the spin-orbital scattering constant of bulk trivial pockets. It is concluded[139] that the relationship between an observed negative magnetoresistance and the supposed existence of a chiral anomaly has to be carefully checked. Because of spin degeneracy lifting, the spin orientations of Weyl fermions can be parallel or antiparallel to the momentum, and this is characterized by the helicity. Conservation of the latter strongly protects the transport of Weyl fermions, which can be effectively scattered only by magnetic impurities. Chemical doping with magnetic and non-magnetic impurities is thus more convincing than is the negative magnetoresistance criterion for detecting the existence of Weyl fermions. Another cautionary tale is offered by a high-field magnetotransport study of an ultra-high mobility n-type GaAs quantum well. A very large linear magnetoresistance, of the order of 100000% was observed[140] in fields of up to 33T, onto which quantum oscillations became superposed in the quantum Hall regime at low temperatures. It was again noted that linear magnetoresistance is often taken to be evidence for the incidence of exotic quasi-particles in materials such as topological semimetals. The observation of such a marked linear magnetoresistance in an ultra-simple system having a free electron-like band structure and an almost defect-free structure eliminated most of the complicated alternative explanations for linear magnetoresistance and suggested density fluctuations were the main cause of the observations. The occurrence of featureless linear magnetoresistance at high temperatures and of quantum oscillations at low temperatures obeyed the empirical resistance-rule that longitudinal conductance is directly related to the derivative of the transversal (Hall) conductance, multiplied by the magnetic field and by a factor which is constant over the entire temperature range. Only at low temperatures were small deviations from the rule observed, and those were more likely to originate from a different transport mechanism for composite fermions. A magneto-optical study of Landau quantization has shown[141] that high magnetic fields drive the system toward the quantum limit, leading to zeroth chiral Landau levels in two inequivalent Weyl nodes. The zeroth chiral Landau level exhibits a clear linear dispersion in the magnetic field direction and permits optical transitions without the limitation of zero z-momentum or magnetic field evolution. The magnetic-field dependence of the zeroth Landau levels also confirms a predicted particle-hole asymmetry of the Weyl cones. Optical transitions from the normal Landau levels reveal the coexistence of multiple carriers. These include a massive Dirac fermion, thus hinting at a more complex topological nature existing in inversion-symmetry breaking Weyl semimetals. First-principles density functional theory calculations, of spin-orbit interactions within the independent-particle approximation, of the family (TaAs, TaP, NbAs, NbP) of body-centered tetragonal topological Weyl semimetals showed[142] that the small energetic overlap of tantalum 5d or niobium 4d conduction and valence bands produced electron and/or hole pockets near to the Fermi energy at the 24 Weyl nodes. Those nodes then gave rise to two-dimensional Dirac cones of Weyl type-I or three-dimensional cones of Weyl type-II. The band dispersion and occupation near to the Weyl nodes governed the infra-red optical properties. Predominant interband transitions led to a deviation from the expected constant values of the imaginary part of the dielectric function. The resultant polarization anisotropy was seen in the real part of the optical conductivity, where the line-shape deviated from the expected linearity. The spectral features of the anisotropic and tilted Weyl fermions were limited to low excitation energies above the onset of absorption due to band occupation. A marked anomaly in the magnetic torque, upon entering the quantum limit state in high magnetic fields, has been reported. The torque changes sign at that limit, indicating a reversal of the magnetic anisotropy which can be directly attributed to the topological nature of the Weyl electrons. It was concluded[143] that such an anomalous quantum-limit torque could provide a direct experimental means for distinguishing Weyl and Dirac systems.

The effect of hydrostatic pressure on magnetotransport was to produce subtle changes in the resistance profile at up to 2.31GPa[144]. The Fermi surfaces exhibited anisotropic changes, with the extremal areas slightly increasing in one plane but decreasing in another. The topological features of the two pockets observed at atmospheric pressure remained unchanged at 2.31GPa. No superconductivity was observed at 0.3K at any of the pressures measured. This agreed with the conclusions of another study[145] in which high-pressure synchrotron X-ray diffraction and resistance measurements had been made. The crystal structure remained stable at up to 26GPa and the resistance of single crystal increased monotonically with pressure at low temperatures. No superconducting transition was observed under pressures of up to 20GPa. It was pointed out that pressure might not be a good means of making a topological superconductor from this Weyl semimetal. Fitting the temperature-dependence of the specific heat to the Debye model, a small Sommerfeld coefficient and a large Debye temperature were deduced; thus confirming that this material is stable under pressure. High-pressure studies of NbAs and TaAs have also revealed[146] that the frequencies of the first-order Raman modes hardened with increasing pressure and exhibited a slope-change at 15GPa in NbAs and 16GPa in TaAs. The resistivities went through a minimum at pressures which were close to the latter transition pressures, and a change in the bulk modulus occurred. First-principles calculations revealed that the transition was associated with an electronic Lifshitz transition in NbAs, but with a structural transition from body-centered tetragonal to hexagonal in TaAs. In this regard, crystal-structure search methods and first-principles calculations have been used[147] to catalogue the high-pressure phase diagrams of NbP, NbAs, TaP and TaAs. This showed that NbAs and TaAs have similar phase diagrams, with the same structural transition sequence: $I4_1md \rightarrow P\bar{6}m2 \rightarrow P2_1/c \rightarrow Pm\bar{3}m$, but slightly different transition pressures. The sequence for NbP and TaP differed slightly, in that new structures appeared, such as Cmcm in NbP and Pmmn in TaP. In the electronic structure of the high-pressure phase, $P\bar{6}m2$, of NbAs it was found that there were coexisting Weyl points and triply-degenerate points; similar to those found in high-pressure $P\bar{6}m2$-TaAs.

TaAs

The discovery of the first Weyl semimetal in TaAs was the first observation of a Weyl fermion in Nature. This topological semimetal features an anomalous surface state, the Fermi arc, which connects a pair of Weyl nodes through the crystal boundary. In the case of the TaAs family, the intrinsic Fermi levels of the Weyl semimetal are not located precisely at the energy of the chiral nodes, and the resultant Fermi surfaces are ellipsoids which enclose the chiral nodes. In the case of Weyl semimetals, the band-crossing points are doubly-degenerate, have a definite chirality and are distributed over an even number of discrete points in the Brillouin zone. This has been theoretically and experimentally proved for the TaAs family of materials. Three-dimensional Dirac semimetals having four-degenerate Dirac nodes near to the Fermi-level can usually lift the spin degeneracy of the energy bands and turn into three-dimensional Weyl semimetals by breaking the time-reversal or inversion symmetries. This again has been theoretically and experimentally confirmed in case of the non-magnetic inversion-symmetry breaking TaAs-type materials. Weyl semimetal states in the TaAs family can be directly detected by means of angle-resolved photo-emission spectroscopy, as evidenced by the existence of Weyl node pairs and linearly dispersed Weyl semimetal bands.

Fermi-arcs can produce three-dimensional quantum Hall effects in topological semimetals such as TaAs, Cd_3As_2 and Na_3Bi, even though that effect is usually restricted to two-dimensional systems, given that topological constraints mean that the Fermi-arc has an open Fermi surface which cannot support the quantum Hall effect. Fermi arcs on opposite surfaces can nevertheless form a complete Fermi loop, via a tunnelling effect assisted by Weyl nodes, and thus enable the quantum Hall effect[148]. The edge states of the Fermi arcs here exhibit a three-dimensional distribution which is very different to the surface-state quantum Hall effect for the single surface of a topological insulator. As the Fermi energy passes through the Weyl nodes, the sheet Hall conductivity changes, from a reciprocal B-dependence, to quantized plateaux at Weyl nodes.

Theoretical calculations were made[149] of the quasi-particle interference patterns that arise from surface states, including topological Fermi arcs in Weyl semimetals like TaAs and NbP. The quasi-particle interference exhibited termination points that are indicative of Weyl nodes in the interference pattern. The results suggested that these were a universal indicator of the existence of topological Fermi arcs in TaAs. Long-sought Weyl nodes were detected in this material by means of bulk-sensitive soft X-ray angle-resolved photo-emission spectroscopy[150]. Projected locations at the nodes on the (001) surface closely matched the Fermi arcs, thus providing definite experimental evidence for the existence of Weyl fermionic quasi-particles. This was important because the massless Dirac equation in three-dimensional momentum space can be regarded as being the overlap of two Weyl fermions of opposite chirality. The Dirac fermionic quasi-particle is stable when protected by a crystal symmetry in topological Dirac semimetals. On the other hand, a separate single Weyl node requires no symmetry protection. Weyl nodes appear in pairs of opposite chirality. An isolated Weyl node is a sink or source of Berry curvature, like a monopole in momentum space (see the description of Fe_3GeTe_2), and the chirality corresponds to its topological charge. In order to obtain isolated Weyl nodes, the spin degeneracy of the electronic bands must be removed by breaking the inversion or time-reversal symmetry. The identification of a material having only Weyl nodes near to the Fermi energy was difficult until the present, non-centrosymmetric and non-magnetic, material was predicted to be a Weyl semimetal having twelve pairs of Weyl nodes in its three-dimensional Brillouin zone. Scanning tunnelling microscopy and spectroscopy have been used to investigate[151] the behavior of electronic states on the surface, and in the bulk, of topological semimetal phases. Upon mapping the quasi-particle interference and emerging Landau levels in high magnetic fields, for the Dirac semimetals, Cd_3As_2 and Na_3Bi, extended Dirac-like bulk electronic bands were identified. Quasi-particle interference imaging of the Weyl semimetal, TaAs, revealed the expected momentum-dependent delocalization of Fermi-arc surface states in the vicinity of surface-projected Weyl nodes. The real and imaginary parts of the dielectric function, and the energy-loss spectra, of TaAs and Na_3Bi were calculated within the random phase approximation[152]. Electron–hole interaction was incorporated by solving the Bethe–Salpeter equation for the electron–hole Green's function. A clear difference was found between the linear optical responses of TaAs and Na_3Bi in the high-energy region where, unlike Na_3Bi, the arsenide exhibited excitonic peaks at 9 and 9.5eV. It was notable that the excitonic effects in the high-energy range of the spectrum were stronger than those in the lower range. The dielectric function was generally red-shifted, as compared with that of the random-phase approximation. The resultant static dielectric constants for Na_3Bi were smaller than the corresponding ones for TaAs. In the low-energy region, the absorption intensity of TaAs was greater than that for Na_3Bi. The calculated second-order non-linear optical susceptibilities showed that the arsenide acted like a Weyl semimetal, and exhibited high values of non-linear response in the low-energy region. In the low-energy region, the optical spectra were dominated by 2ω intra-band contributions. The TaAs family is considered[153] to be the ideal class of materials class for exploring the expected effects of Weyl points: from Fermi arcs to chiral magnetotransport. It is to be noted that recent structural investigations have revealed[154] the presence of an hexagonal phase with a WC-type P6m2 structure, intergrown with the known tetragonal form of this material. The hexagonal c-axis lay along the $[100]_{tetra}$ or $[010]_{tetra}$ direction.

TaAs₂

This non-magnetic semimetal exhibits a very large negative magnetoresistance which is associated with an unknown scattering mechanism. Density functional calculations showed[155] that it is a topological Z_2 semimetal, (0;111), without Dirac dispersion. This demonstrated that the negative magnetoresistance of a non-magnetic semimetal cannot be definitely attributed to the Adler-Bell-Jackiw chiral anomaly of bulk Dirac/Weyl fermions. The resistivity in a magnetic field of high-quality monocrystals of $NbAs_2$ and $TaAs_2$ having inversion symmetry has been investigated[156]. Both of them exhibited metallic behavior in a zero magnetic field, and a metal-to-insulator transition occurred when a magnetic field was applied. This led to an extreme magnetoresistance of 100000% in the case of $NbAs_2$, and of 730000% in the case of $TaAs_2$ at 2.5K in a field of 14T. With decreasing temperature, a resistivity plateau appeared beyond the insulator-like range.

An unsaturated transverse extreme magnetoresistance of about 300000% at 2K and 15T, with a quadratic field dependence, has been observed[157] in $NbAs_2$ single crystals. At up to 12.5K, clear Shubnikov de Haas oscillations were observed, which helped to identify two distinct Fermi pockets. From the field-dependent Hall resistivity at 2K, carrier concentrations of 6.7691 x 10^{25} and 6.4352 x $10^{25}/m^3$ were deduced for electrons and holes, respectively, while the corresponding mobilities were 5.6676 and 7.6947m^2/Vs. An anisotropic magnetoresistance of about 84, 75 and 12% was found at 0.75T and 6K for various orientations, like that found for some topological semimetals. The magnetic properties of NbAs$_2$ were similar to the case of graphite, with no phase transition between 5 and 300K.

TiRhAs

First-principles calculations[158] have predicted that this material is a Dirac nodal line semimetal, with the Dirac nodal line being protected by a combination of inversion, time-reversal and reflection symmetries, in the absence of spin-orbital coupling. Calculations showed that the band velocities which were associated with the nodal line exhibited a high degree of directional anisotropy, with the in-plane velocities perpendicular to the nodal line being between 1.2 x 10^5 and 2.8 x 10^{5}m/s. Crossings along the Dirac nodal line also exhibited a prominent position-dependent tilt along the directions perpendicular to the nodal line. The Z$_2$ indices, whose values were based upon parity eigenvalues at time-reversal invariant momenta, showed that the material is topological. A tight-binding model indicated the existence of two-dimensional drum-head surface states on the surface Brillouin zone.

W$_2$As$_3$

As well as Dirac semimetals, Weyl semimetals, nodal-line semimetals and semimetals having triply-degenerate nodal points, there are Z$_2$ topological metals; typified by non-trivial Z$_2$ topological invariants, topological surface states and the absence of a bulk energy gap. These are expected to include LaP, LaAs, LaSb and LaBi, with the latter being the only one to exhibit topological non-trivial band dispersion, as proved by means of angle-resolved photo-emission spectroscopy. The present compound is a topological semimetal with a C2/m space group. First-principles calculations show that the band crossings are partially gapped if spin-orbital coupling is included. The Z$_2$ indices of electron filling are such as to indicate that it is a marked topological insulator with topological surface states. Magnetotransport measurements, and a nearly quadratic field-dependence of magnetoresistance at 3K, with the field parallel to [200], indicate it to be an electron-hole compensated compound in which the longitudinal magnetoresistance reaches 11500% at 3K in a field of 15T[159]. Multi-band features are revealed by high magnetic-field Shubnikov-de Haas oscillation, Hall resistivity and band calculations. A non-trivial π Berry phase is predicted, reflecting the proposed topological features of the material. A two-band model closely fits the conductivity and Hall-coefficient data.

Bismuthides

GdPtBi

Many half-Heusler compounds of the form, RPtBi, are predicted to be topological semimetals and thus exhibit topological surface states, a chiral anomaly and planar Hall effect. An unusual intrinsic anomalous Hall effect has been observed[160] in the antiferromagnetic Heusler Weyl semimetals, GdPtBi and NdPtBi, over a wide range of temperatures. The GdPtBi composition has an anomalous Hall conductivity of up to 60S/cm and an anomalous Hall angle which can be as large as 23%. Muon spin-resonance data indicate the existence of a sharp antiferromagnetic transition at 9K, with no appreciable magnetic correlations at higher temperatures. It is deduced that Weyl points are induced in these half-Heusler alloys, by a magnetic field, via exchange splitting of the electronic bands at, or near to, the Fermi energy and is the source of the chiral anomaly and anomalous Hall effect.

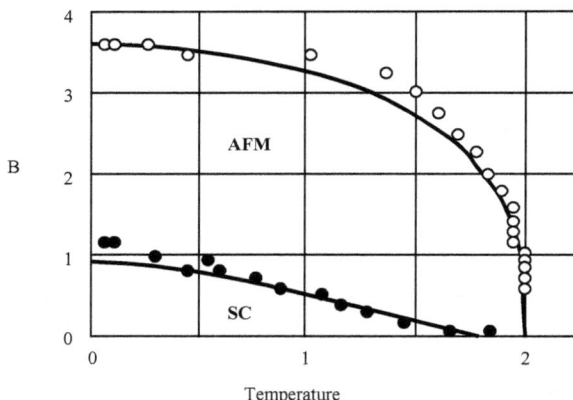

Figure 11. Magnetic phase diagram of HoPdBi
AFML antiferromagnetic, SC: superconducting

HoPdBi

Magnetotransport studies[161] of the magnetic and electronic properties of this non-centrosymmetric half-Heusler antiferromagnetic (T_N = 2K) alloy have revealed semimetallic behavior, with a carrier concentration of $3.7 \times 10^{18}/cm^3$; as deduced from the Shubnikov-de Haas effect. The magnetic phase diagram (figure 11) was determined by means of transport, magnetization and thermal expansion measurements. As the temperature approaches 0 in a field of 3.6T, the magnetic order is suppressed. Superconduction with a critical temperature of 1.9K is found in the antiferromagnetic phase. The upper critical field exhibit an unusual linear temperature variation. Electronic-structure calculations show that the material is a topological semimetal which exhibits a non-trivial band inversion of 0.25eV.

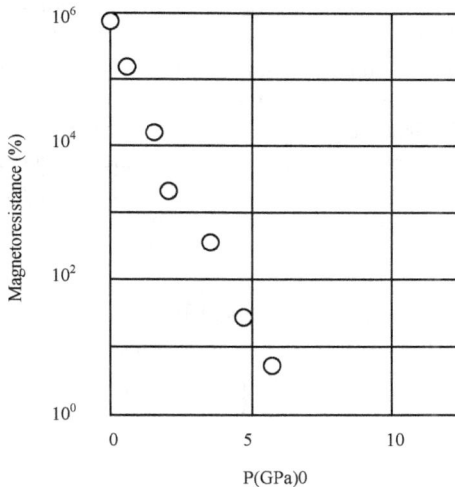

Figure 12. Magnetoresistance (9T, 2K) of LaBi as a function of pressure

The specific heat exhibits anomalies which correspond to an antiferromagnetic ordering transition and crystalline field effect, but not to a superconducting transition. Single-crystal neutron diffraction indicates[162] that the antiferromagnetic structure is characterized by the propagation vector. The temperature dependence of the electrical resistivity involves two parallel conducting channels, having semiconducting and metallic natures. In a weak magnetic field, the magnetoresistance exhibits a weak antilocalization effect. In a strong field, and at temperatures below 50K, it is large and negative. Below 7K, Shubnikov-de Haas oscillations having two frequencies appear - in the resistivity - which have a non-trivial Berry phase; a typical signature of Dirac fermions. In this family of topological semimetals, RPdBi; ternary half-Heusler compounds in which R is a rare-earth element, a suitable choice of the rare-earth f-electron component permits the simultaneous modification[163] of the lattice density, via the lanthanide-contraction effect, and of the strength of the magnetic interaction, via de Gennes scaling. This thereby provides an unique opportunity to tailor the normal-state band inversion strength, superconducting pairing and magnetically ordered ground states. Antiferromagnetism with the ordering vector, (½,½,½), occurs below a Néel temperature that scales with the de Gennes factor. Any superconducting transition is simultaneously suppressed by an increased de Gennes factor. In general, monocrystals of equiatomic ternary compounds of rare-earth elements with palladium and bismuth have a cubic MgAgAs-type half-Heusler structure. Band-structure calculations show[164] that many members of the family exhibit electronic band inversion, leading to topological insulator or topological semimetal behaviour. Even in the absence of intrinsic band inversion, a topologically non-trivial state can theoretically occur via a specific antiferromagnetic order. The antiferromagnetic structures of all of the bismuthides studied so far are characterized by the propagation vector, allowing for an antiferromagnetic topological insulator state.

InBi

This material has a simple tetragonal layered structure, with a P4/nmm space group, lattice constants of a = 5.014Å and c = 4.784Å and a cleavage plane along the [001] direction which is due to relatively weak bonding in the neighboring bismuth sub-layer. The Brillouin zone is square, with four-fold symmetry. Investigation[165] of the bulk and surface electronic structures of this non-symmorphic material, which is also a type-II Dirac semimetal, confirmed it to be a topological nodal line semimetal. Given the strong spin-orbital coupling in the material, the persistence of nodal lines demonstrates the great protection afforded by the non-symmorphic symmetry of the crystal structure. In two-band systems, non-symmorphic symmetries (e.g. unitary non-symmorphic, non-symmorphic magnetic symmetry, non-symmorphic symmetry combined with inversion) can unavoidably create band-crossings in the bulk, and lead to reduced-dimensionality Fermi surfaces[166]. These forced crossings arise from the momentum dependence of the non-symmorphic symmetry by placing severe restrictions on the global structure of any band configuration. In the case of non-symmorphic magnetic symmetry and non-symmorphic symmetry combined with inversion, the band-crossings are located at high-symmetry points of the Brillouin zone; the exact positions being governed by the mathematical details of the symmetry operators.

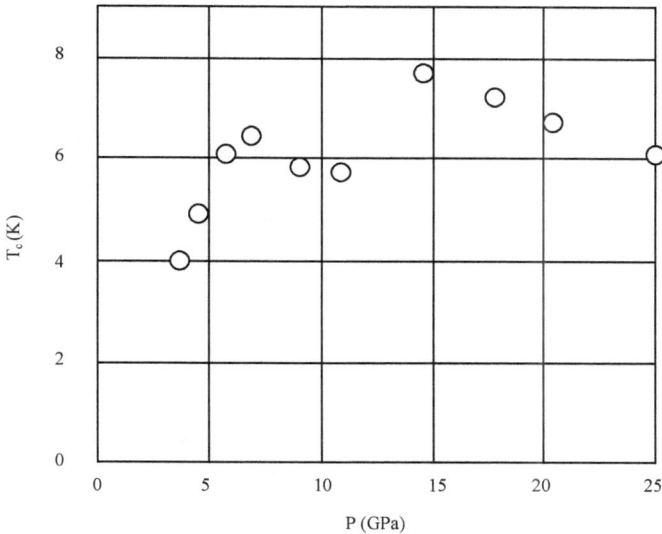

*Figure 13. Temperature-pressure phase diagram of superconductivity in LaBi
The discontinuity is due to a structural transition from cubic to tetragonal*

KMgBi

First-principles calculations predicted[167] that materials of the form, AMgBi, where A is potassium, rubidium or cesium, are symmetry-protected topological semimetals near to the boundary of type-I and type-II Dirac semimetal phases. They are therefore termed topological critical Dirac semimetals. In-plane compressive straining of KMgBi, or doping with rubidium or cesium, can drive transitions between the above two phases. It was possible to describe mathematically the bands near to the Fermi energy, and calculate the surface Fermi-arcs and Landau-levels during transitions. On the other hand, when KMgBi crystals were grown by using the bismuth flux method, they exhibited[168] semiconducting behavior with a metal-semiconductor transition at high temperatures and both electron- and hole-type carriers existed having a strong temperature-dependence of the carrier concentrations and mobilities. This suggested that KMgBi is a narrow-band semiconductor with multiband features in the bulk rather than a semimetal, as predicted theoretically. It also exhibited a resistivity plateau at low temperatures, like a topological insulator, implying that a non-trivial topological surface state might exist in KMgBi.

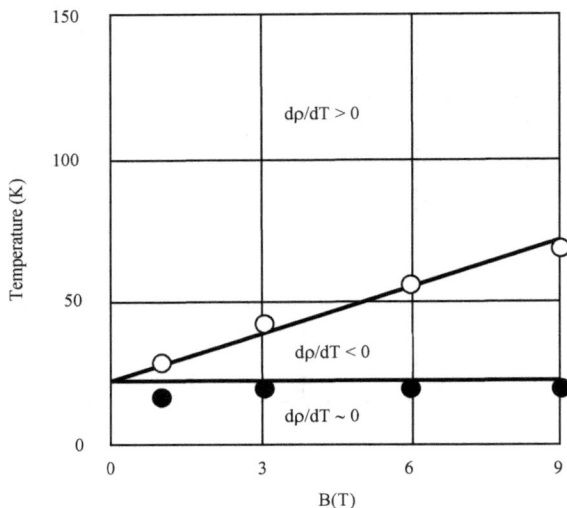

Figure 14. Temperature-field phase diagram of LaBi

LaBi

In order to establish a relationship between extreme magnetoresistance and superconductivity, the effect of pressure upon this material was studied[169]. With increasing pressure, there was a disappearance of extreme magnetoresistance (figure 12) which was followed by the appearance of superconductivity at 3.5GPa (figure 13). There was a region of overlap between the superconductivity and extreme magnetoresistance. Suppression of the latter was associated with an increasing zero-field resistance rather than a decreasing in-field resistance. The suppression of extreme magnetoresistance is also associated with a change in sign of the Hall coefficient from negative to positive. Density functional theory calculations showed that the sign change is due to a shrinkage of the electron pocket with increasing pressure. At a pressure of 11GPa, there was a structural transition from face-centered cubic to a primitive tetragonal, as predicted by theory. The change in crystal structure changes the band-structure and creates a region of band inversion. The changes in the band structure due to structural transition produce a reversal of the 10K/300K resistivity ratio from decreasing to increasing, and a fall in the Hall coefficient. There is also a discontinuity in the critical temperature and the H_{c2}/T_c ratio.

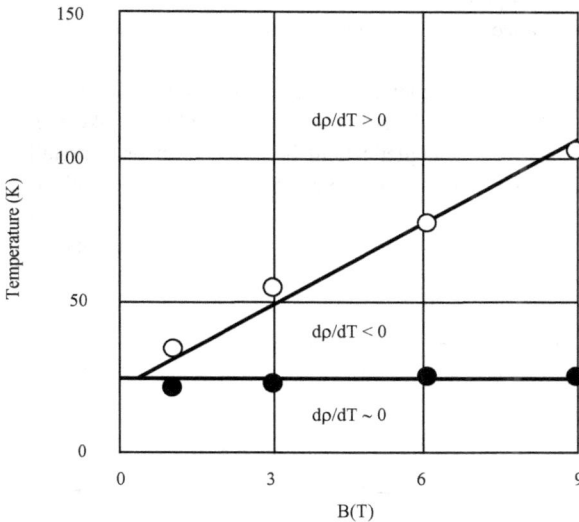

Figure 15. Temperature-field phase diagram of LaSb

A comparative study[170] of magnetotransport effects in LaBi and LaSb yielded temperature-field phase diagrams (figures 14 and 15) which summarise how a magnetic field can modify the electronic behavior of these materials. Similar phase diagrams exist for other topological semimetals; even those having other crystal structures and chemical compositions (figures 16 and 17). It was suggested that extreme magnetoresistance in LaBi and LaSb arises from a combination of compensated electron-hole pockets and the orbital texture of the electron pocket. The layer texture was expected to be a typical feature of various topological semimetals and to produce their low residual resistivity in a zero magnetic field.

High-resolution laser-based angle-resolved photo-emission spectroscopy measurements and density functional theory calculations indicated that both bulk and surface bands were present in the spectra, and that the dispersion of the surface state was very unusual. The latter resembled a Dirac cone, but an energy-gap could be detected[171]. The bottom band exhibited an approximately parabolic dispersion. The dispersion of the top band remained entirely linear; being V-shaped, and with the tip very closely approaching the extrapolated location of the Dirac point. In this semimetal, which has a band inversion that is equivalent to a topological insulator, a surface-state like behavior was observed[172] in the magnetoresistance. The electrons which were responsible for the pseudo two-dimensional transport originated from bulk states rather than from topological surface states; as indicated by angle-dependent quantum oscillations of the magnetoresistance and by *ab initio* calculations. Cigar-shaped electron valleys make the charge transport highly anisotropic when the magnetic field is varied from one crystallographic axis to another. The electrons could be effectively polarized in these electron valleys in a rotating magnetic field. A polarization of 60% was obtained[173] at 2K in spite of the coexistence of three-dimensional hole pockets. The magnetoresistance exhibited a marked anisotropy having an amplitude of 10^5%.

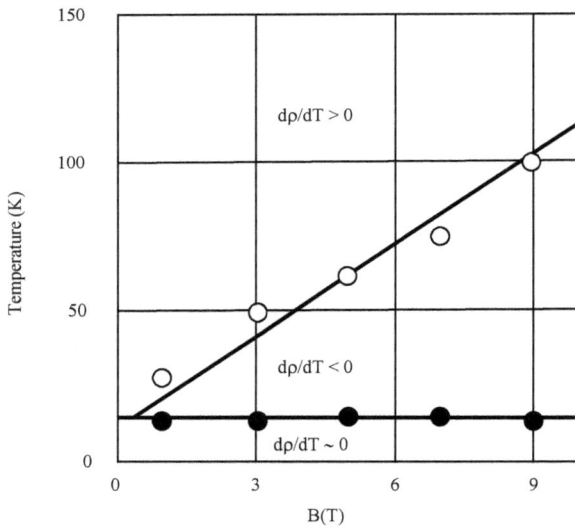

Figure 16. Temperature-field phase diagram of NbSb$_2$

Materials Research Forum LLC
doi: http://dx.doi.org/10.21741/9781644900154

Na₃Bi

Atomically flat thin films of the topological semimetal were grown onto double-layer graphene which had been formed on 6H-SiC(00•1) substrates by means of molecular beam epitaxy, and the band structure of Na_3Bi was mapped along the Γ-M and Γ-K directions. The energy band at higher energies was determined by doping cesium atoms into the surface. First-principles calculations[174] identified a structural phase transition of Na_3Bi from the hexagonal ground-state to the cubic *cF*16 phase at above 0.8GPa, in agreement with experiment. Upon removing the pressure, the *cF*16 phase was mechanically stable under ambient conditions. Calculation showed that the *cF*16 phase is a topological semimetal, similar to HgTe, and has an unusually low C' modulus of about 1.9GPa (table 2) together with a large A^u anisotropy of up to 11. This makes the *cF*16-type phase very soft, with a liquid-like elastic deformation behaviour on the $(110)(1\bar{1}0)$ slip system. This deformation was associated with a topological phase transition from a topological semimetal state, in its strain-free cubic phase, to a topological insulator in the distorted phase.

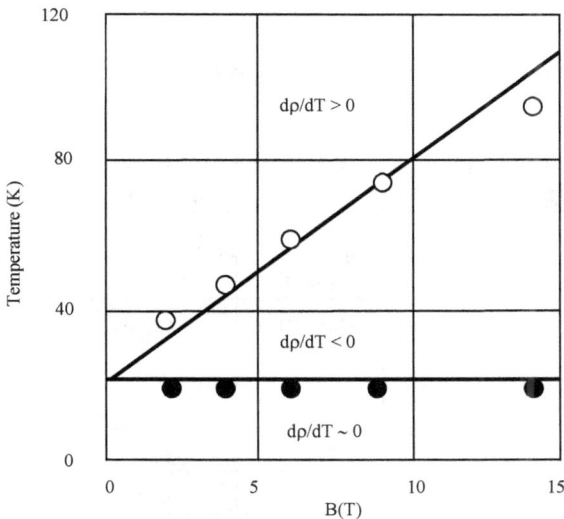

Figure 17. Temperature-field phase diagram of WTe₂

First-principles studies have been made[175] of the electronic properties and transitions of four structural forms, $P6_3/mmc$, $P\bar{3}c1$, $Fm\bar{3}m$ and $Cmcm$, of this material under constant-volume uniaxial strain. In the case of $P6_3/mmc$ and $P\bar{3}c1$, there was a transition from topological Dirac semimetal to topological insulator at a compressive strain of about 4.5%. The insulator gap increased with increasing compressive strain, and reached about 0.1eV at a strain of 10%. The $P63/mmc$ and $P\bar{3}c1$ forms retained the properties of a topological Dirac semimetal under tensile strains of 0 to 10%, although the Dirac points moved away from the Γ-point and along Γ-A in reciprocal space with increasing tensile strain. The $Fm\bar{3}m$ form was a topological semimetal with inverted bands between sodium 3s and bismuth 6p, with parabolic dispersion in the vicinity of the Γ-point. In the case of the $Fm\bar{3}m$ form, compressive and tensile strains both led to a topological Dirac semimetal which was identified by calculating the surface Fermi arcs and topological invariants at time-reversal planes in reciprocal space. The $Cmcm$ high-pressure form was an ordinary insulator, with a gap of about 0.62eV. The gap remained almost constant at about 0.60eV under compressive strains of 0 to 10%. A transition from insulator to metallic phase occurred, under tensile straining, which was highly sensitive and occurred only at a strain of 1.0%.

Table 2. Calculated elastic constants of Na_3Bi, compared with those of HgTe

Constant	Na_3Bi	HgTe
C_{11}	22.3GPa	43.0GPa
C_{12}	18.4GPa	28.7GPa
C_{44}	21.9GPa	17.2GPa
C'	1.950GPa	7.187GPa
E	23.8GPa	32.5GPa
A	11.231	2.397
Chung ratio	0.524	0.089
Universal ratio	11.020	0.977

This material is a prime example of a Dirac semimetal which is protected by rotational symmetry. Topological states are often protected by a lattice symmetry. In three dimensions, the Berry flux around gapless excitations in momentum space define a chirality, thus a protective symmetry can also be termed a chiral symmetry. The interplay between the chiral symmetry order parameter and instantaneous long-range Coulomb interaction has been investigated[176] by using standard renormalization-group methods. A topological transition, associated with chiral symmetry, is found to be stable in the presence of Coulomb interactions. The electron velocity always becomes greater than that for a chiral symmetry order parameter. The transition cannot be relativistic, meaning that supersymmetry is inherently forbidden by long-range Coulomb interactions. In order to investigate the breaking of inversion-symmetry related optical properties, a study was made[177] of the optical properties of the present Dirac semimetal, and of BaPt, in comparison with those of the Weyl semimetals, NbP and $Na_3Bi_{0.75}Sb_{0.25}$. The real and imaginary parts of the dielectric function, and the energy-loss spectra, of a topological semimetal - with and without inversion symmetry – were calculated within the random phase approximation. The electron-hole interaction was included by solving the Bethe-Salpeter equation for the electron-hole Green's function. The lack of inversion symmetry and spin-orbital interaction increased the density of states at the Fermi level and produced an excitonic peak in the optical absorption of a given topological semimetal. The excitonic effects in the high-energy range of the spectrum are stronger moreover than in the lower range. The calculations showed that NbP, with no inversion symmetry, has high-energy excitons at 10 and 10.8eV. In contrast to Na_3Bi, electron-hole interactions give rise to several weak peaks of differing energy in the optical absorption of $Na_3Bi_{0.75}Sb_{0.25}$ but its red-shift is less marked.

Finally, a marked chiral-anomaly induced negative magnetoresistance is observed in samples having a low (~10^{17}/cm^3) carrier density. A large linear magnetoresistance is observed in material possessing a high (~10^{18}/cm^3) carrier density.

PdBi$_2$

A spectroscopic and scanning tunnelling microscopic study[178] was made of this centrosymmetric superconductor, which exhibits a topological surface state. A combination of first-principles electronic-structure calculations and quasi-particle interference observations elucidated the spin textures at the surface. This revealed not only the topological surface state, but also all of the other surface bands exhibiting spin polarisation parallel to the surface. The superconducting gap (figure 18) was fully open in all of the spin-polarised surface states. This behaviour was consistent with the spin-triplet order parameter that was expected for such in-plane spin textures. The observed superconducting gap amplitude was however comparable to that of the bulk, thus suggesting that the spin-singlet component predominated in β-PdBi$_2$.

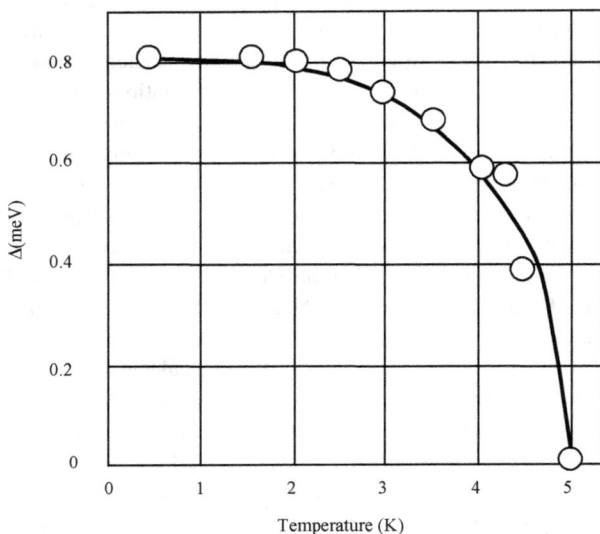

Figure 18. Temperature dependence of the superconducting gap of PdBi₂

PtBi₂

Angle-dependent magnetoresistance measurements of monocrystals of this face-centered cubic pyrite-type material, with a space group of Pa$\bar{3}$ and lattice parameter of 6.702Å, in high magnetic fields revealed unsaturated magnetoresistances of up to 11.2×10^{6}% at 1.8K in a magnetic field of 33T. The crystals exhibited ultra-high mobility and large Shubnikov-de Hass quantum oscillations, together with a non-trivial Berry phase[179]. Analysis of the Hall resistivity suggested that the extreme magnetoresistance could be attributed to nearly-compensated electrons and holes. The results implied that this material is a possible topological semimetal.

Table 3. Lattice parameters of $Gd_xY_{1-x}PtBi$

x	Lattice Parameter (Å)
0	6.76
0.25	6.78
0.50	6.79
0.75	6.80
1	6.81

YPtBi

This non-centrosymmetric half-Heusler compound was found[180] to be superconducting below a critical temperature of 0.77K, with a zero-temperature upper critical field of 1.5T. Magnetoresistance and Hall measurements were in accord with theoretical predictions that it is a topologically non-trivial semimetal with a positive charge-carrier density of 2 x 10^{18}/cm^3. An unusual linear magnetoresistance, and beats in Shubnikov-de Haas oscillations, suggested the existence of spin-orbital split Fermi surfaces. A sensitivity of the magnetoresistance to surface roughness suggested that surface states perhaps played a role. A nearly vanishing density of states around the Fermi level would normally imply vanishing electron-phonon coupling and therefore not permit superconduction. On the basis of relativistic density-functional theory calculations of electron-phonon coupling, it was concluded[181] that carrier concentrations greater than 10^{21}/cm^3 would be required in order to explain the observed critical temperature using a conventional pairing mechanism. It is thus probable that an unconventional pairing mechanism is responsible for the superconduction in and related topological semimetals having the half-Heusler structure. It was recalled[182] that, in fermionic superfluids, Cooper pairs comprise spin-½ quasi-particles which pair to form spin-singlet or spin-triplet bound states. The spin of a Bloch electron is instead fixed by the crystal symmetry and the atomic orbitals, such that it can sometimes behave like a spin-3/2 particle. The superconducting state of such systems permits pairing to occur beyond spin-triplet, and higher-spin quasi-particles can combine to form quintet or septet pairs. Unconventional superconduction is thought to arise from a spin-3/2 quasi-particle electronic structure in this half-Heusler semimetal: a low carrier-density non-centrosymmetric cubic material of high symmetry which preserves the p-like j = 3/2 manifold in the bismuth-based G8 band in the presence of strong spin-orbit coupling. The existence of line nodes in the superconducting order parameter is then directly explainable in terms of a mixed-parity Cooper pairing model with a high total angular momentum; consistent with a high-spin fermionic superfluid state.

Density functional theory and generalized gradient approximation studies have been made[183] of the topological phases of $Gd_xY_{1-x}PtBi$ (x = 0, 0.25, 0.5, 0.75, 1) compositions. Apart from the case of x = 0, the alloys are ferromagnetic under normal conditions. The lattice constants (table 3) and bulk moduli (table 4) are linear or non-linear functions of x, respectively. Calculated electron densities of state show that the alloys are metallic, with a low conductivity. The coefficient of linear electronic specific heat could be fitted by a third-order polynomial. For all x-values, the materials have an inverted band order and are possible topological semimetals. The band-inversion strength exhibits a non-linear x-dependence. The magnetic moment increases with increasing gadolinium content, and is a linear function of x.

Table 4. Bulk moduli of $Gd_xY_{1-x}PtBi$

x	Modulus (GPa)
0	86.04
0.25	86.98
0.50	87.27
0.75	86.50
1	83.17

First-principles calculations, combined with Boltzmann transport theory, of the thermoelectric properties of half-Heusler compounds of the form, MPtBi, where M is scandium, yttrium or lanthanum, showed[184] that they are all topological semimetals at equilibrium, but become topological insulators under uniaxial tensile strain. It was predicted that a thermoelectric performance which was comparable to that of Bi_2Te_3 could be achieved by the half-Heusler topological insulator, LaPtBi, upon applying a tensile uniaxial strain of 8%. It was again confirmed that the lattice thermal conductivity of LaPtBi is anomalously low when compared with that of traditional half-Heusler compounds which are not topological insulators.

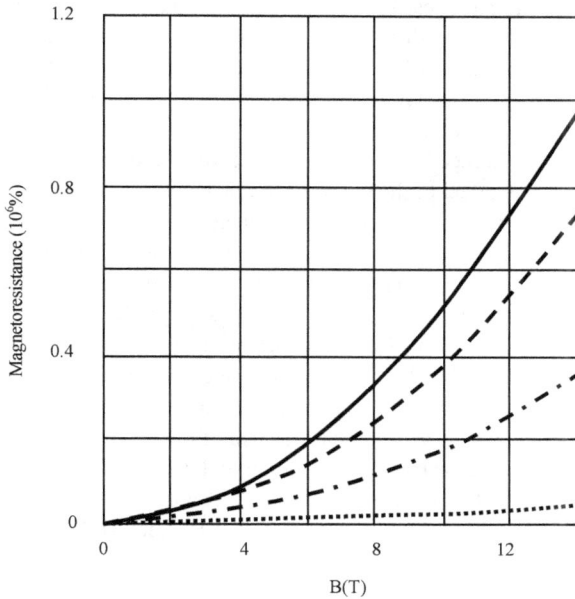

Figure 19. Field dependence of the magnetoresistance of ZrB$_2$
Solid line: 2K, dashed line: 14K, dash-dot line: 40K, dotted line: 60K

YbPtBi

Electronic structure calculations and angle-resolved photo-emission spectroscopic measurements indicate that, at high temperatures, pairs of triply degenerate points exist in this material. In an external magnetic field, these points split into pairs of Weyl nodes and the presence of Weyl fermions is revealed by angle-dependent magnetotransport data. At low temperatures, where the electronic structure is strongly affected by band hybridization, heat-capacity measurements have suggested the occurrence of nodal thermal excitation; considered to be evidence for the presence of a Weyl Kondo semimetal phase. This is also supported by the detection of a topological Hall-effect in relevant resistivity measurements. A relationship between carrier density and negative longitudinal magnetoresistance[185] strongly suggests the presence of a chiral anomaly and can be explained on the basis of the band structure. A clear peak in the Hall resistivity in a near-zero field appeared at the point where an electron pocket began to emerge, in samples which exhibited the strongest signs of a chiral anomaly. Upon increasing the chemical potential, the hole pocket shrank continuously. At a certain point, the Fermi level crossed the electron pocket. Further upward shifts of the Fermi level resulted in a continuous increase in the electron-type carrier density. This was consistent with Hall resistivity measurements. The strength of the chiral anomaly meanwhile depended strongly upon the energy difference between the Fermi level and the Weyl nodes. When they were closer together, the effect of the nodes upon magnetotransport was greater. The association of the strongest negative longitudinal magnetoresistance with electron-band appearance suggested that the Weyl nodes are located close to the bottom of the electron band. It was cautioned that there are other possible reasons for the appearance of negative longitudinal magnetoresistance, especially in the case of a magnetic system such as the present one, because the suppression of inelastic scattering upon applying a magnetic field could also lead to a negative longitudinal magnetoresistance. In that case, the strength of the negative magnetoresistance is not expected to depend greatly upon the carrier density. But since the negative magnetoresistance of the present material exhibits an appreciable dependence on the carrier concentration and on the position of the Fermi level, this implied that the effect here arose from the chiral anomaly. Analysis of the Hall resistivity data reveals clear signs of an anomalous Hall effect at low temperature; possibly related to a complex Berry curvature in momentum space.

Boron and Borides

Borophene

This material exhibits some notable properties, such as superconductivity, and topological features[186]. A tight-binding model of 8-Pmmn borophene was developed[187] which confirmed that the crystal features massless Dirac fermions and that the Dirac points are protected by symmetry. When strain was introduced into the model, it induced a pseudomagnetic field vector potential and a scalar potential. A theoretical investigation[188] of the intrinsic carrier mobility in semimetals having tilted Dirac cones under longitudinal and transverse acoustic phonon scattering led to an analytical formula for the carrier mobility which showed that tilting greatly reduced mobility. The theory was applied to 8B-Pmmn borophene and to its fully hydrogenated form (borophane), both of which have tilted Dirac cones. The predicted carrier mobilities in the borophene at room temperature were 14.8×10^5 and $28.4 \times 10^5 cm^2/Vs$ along the x- and y-directions, respectively. The borophane, in spite of its super-high Fermi velocity, had a carrier mobility which was lower than that of the borophene, due to its lower elastic constant under shear strain. The topological semimetal, bilayer hexagonal borophene, is a graphene analogue consisting of two layers linked by pillars. Density functional theory calculations show[189] that it is a Dirac material which exhibits a nodal line.

The various possible non-magnetic topological semimetals and their k•p Hamiltonian for all layer groups having multiple screw axes, in the absence of spin-orbital coupling, have been enumerated[190]. A so-called cat's-cradle like topological semimetal phase which resembled a staggered multiple hourglass-like band structure was found. A pair of tilted anisotropic Dirac-cones at the Fermi level was expected to occur in two-dimensional boron-based materials such as borophene and borophane. As in the case of a three-dimensional Weyl semimetal, the topological nature of such cat's-cradle Dirac semimetals could be confirmed by calculating the quantized Berry phase and a flat Fermi-arc edge state connecting two Dirac points.

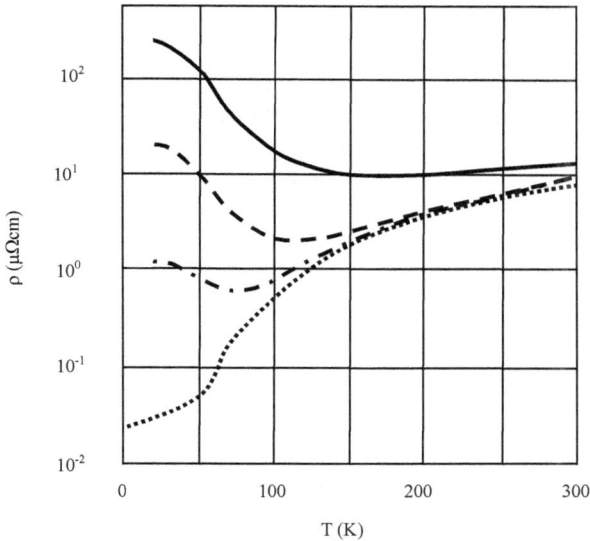

Figure 20. Temperature dependence of the resistivity of ZrB_2
Solid line: 14T, dashed line: 4T, dash-dot line: 1T, dotted line: 0T

A tight-binding model, based upon the Slater-Koster method, accurately reproduces the electronic spectrum and an effective four-band model Hamiltonian describes the spectrum near to the nodal line. That Hamiltonian could be used to study the properties of nodal-line semimetals. Such a nodal line is created by edge states and resists the effects of perturbations and impurities. It was noted that breaking symmetries could split the nodal line but could not open a gap. The β12 polymorph of borophene, when grown on Ag(111) surfaces, also features Dirac fermions[191]. Like graphene, this form of borophene can be viewed as being an atom-vacancy pseudo-alloy on a close-packed triangular lattice. Here however, the origin of the Dirac fermions remains unclear. First-principles calculations have shown that free-standing sheets harbour two Dirac cones and a topologically non-trivial Dirac nodal line with Dirac-like edge states. When grown on Ag(111), the Dirac cones develop a gap while the topologically protected nodal line remains intact and its position in the Brillouin zone agrees with the Dirac-like electronic states detected by experiment. This again confirms the accuracy of computational approaches and their useful guidance in the experimental exploration of two-dimensional materials[192]. There has been little study of its magnetoresistance[193].

ZrB$_2$

Study of the transport properties of single crystals of this predicted topological nodal-line semimetal reveals[194] an extremely high magnetoresistance (figure 19) as well as a field-induced resistivity (figure 20). These features can be explained in terms of a two-band model with perfect electron-hole compensation and high carrier mobilities. Electrons having small effective masses and non-trivial Berry phase are present in high densities as compared with those in other topological semimetals, thus suggesting that this material contains Dirac-like nodal-line fermions.

Carbon and Carbides

C

Rhombohedral graphite can serve as a generic model for a topological semimetal which exhibits an interaction-driven transition on its surface. Fluctuations of the amplitude mode in a superconducting system of coupled Dirac particles was proposed[195] as being a possible model for surface or interface superconductivity. The absence of Fermi energy, and vanishing of the excitation-gap of the collective amplitude mode in the model, led to a large fluctuation-contribution to thermodynamic properties such as the heat capacity. Mean-field theory here became inaccurate, thus indicating that the interactions led to a strongly correlated state. In addition to the usual Weyl points and Dirac lines of three-dimensional topological semimetals, a more complicated momentum-space topological defect has been described[196] in which several separate Dirac lines interconnect to form a momentum-space equivalent of a real-space nexus which had previously been considered

Topological Semimetals   Materials Research Forum LLC
Materials Research Foundations **48** (2019)   doi: http://dx.doi.org/10.21741/9781644900154

for helium-3. When close to the nexus, the Dirac lines were expected to undergo a transition from type-I to type-II. In a general stacked honeycomb lattice model having the symmetry of Bernal stacked graphite, the structural mirror symmetries protected the existence of Dirac lines and naturally led to nexus formation. Due to the bulk-boundary correspondence of topological media, the presence of the Dirac lines then led to the formation of drum-head surface states on the side-surfaces. First-principles calculations[197] have identified a carbon allotrope having a simple orthorhombic crystal structure of Pbcm symmetry. It can be constructed by inserting zig-zag carbon chains between the graphene layers in graphite, or by crystalline modification of a (3,3) carbon nanotube using a double-cell reconstruction mechanism. Dynamic stability was confirmed by performing phonon and molecular dynamics simulations, and electron band calculations indicated that it is a nodal-line semimetal with two nodal lines that pass right through the entire Brillouin zone in the bulk, and with a projected surface flat band around the Fermi level.

First-principles calculations have identified[198] a novel topological semimetal carbon phase in all-sp^2 networks having a 16-atom body-centered orthorhombic unit cell which is comparable to solid face-centred cubic C_{60} with regard to energetic stability. This carbon allotrope is regarded as being a three-dimensional modification of graphite. Simulated X-ray diffraction patterns closely matched previously unexplained diffraction peaks which had been seen in the X-ray diffraction spectra of detonation products and chimney soot. Electronic band structure calculations reveal that C_{16} is a topological node-line semimetal having a single nodal ring. Further research has shown[199] that body-centred tetragonal C_{16} is a topological node-line semimetallic carbon phase which is composed of tetra-rings. Elastic constant calculations and molecular dynamic simulations show that it is energetically metastable, with respect to graphite and diamond, by 0.649 and 0.551eV/atom, respectively. This is attributed to a higher ring-strain in tetra-rings.

The possibility of using a carbon-based porous topological semimetal as a lithium-battery anode material was analyzed[200] by using density-functional theory and cluster-expansion methods. The topological semimetal, bco-C16, is a promising anode material which has a higher specific capacity than that of commonly used graphite anodes. The lithium ions in bco-C16 exhibit one-dimensional migration and the ion-diffusion channels resist the compressive and tensile strains which occur during charging and discharging. The energy barrier decreases with increasing lithium insertion and can attain 0.019eV at high lithium-ion concentrations. The average voltage can be as low as 0.23V, and the volume change during use is comparable to that of graphite.

B$_2$C

The intrinsic two-dimensional carbide possesses a slightly buckled configuration of boron and carbon sub-lattices having a vertical separation of 0.025Å. The unit-cell constants are 2.58 and 3.42Å for the x and y directions, respectively. The space group is Pmm2, with the point group, C$_{2v}$, in which only a two-fold rotation along the z-axis and two mirror symmetries perpendicular to the x and y axes exist. The carbon atom receives 0.1|e| from two equivalent boron atoms. Density functional theory shows[201] that nodal line semimetals and Dirac semimetals can co-exist in the low-energy electronics of a two-dimensional B$_2$C monolayer. As well as type-I and type-II Dirac fermions, a form of open nodal line can appear around the Fermi level. The low-energy electronics of the B$_2$C sheet can be described in terms of a tight-binding model that relies on a basis of boron p$_y$,p$_z$ and carbon p$_y$,p$_z$ states. The energy windows of these four types of topological semimetal are different and can be easily distinguished by experiment.

MoC

It has been proposed[202] that the hexagonal form of the carbide is a new type of topological semimetal which comprises a complicated Fermi surface consisting of two concentric nodal rings in the presence of spin-orbital coupling, and four pairs of triply-degenerate points in the vicinity of the Fermi energy. The appearance of the triply degenerate crossing point is protected by rotation and mirror symmetries. The coexistence of the nodal ring Fermi surface and triply-degenerate points leads to features such as distinguishable drum-head surface states and adjustable new fermions.

Mo$_2$TiC$_2$

This material is a member of the family of two-dimensional transition metal carbides known as MXenes, some of which are of the form, M$'_2$M$''$C$_2$, where M$'$ and M$''$ are transition metals with M$'$ typically being molybdenum or tungsten and M$''$ being titanium, zirconium or hafnium. Non-trivial topological states of the MXenes are revealed by the Z$_2$ index; based upon the parities of the occupied bands below the Fermi energy, time-reversal invariant momenta and the presence of edge states. MXene oxides, M$'_2$M$''$C$_2$O$_2$ have non-trivial gaps ranging from 0.041 to 0.285eV within the generalized gradient approximation and from 0.119 to 0.409eV within the hybrid functional[203]. The band-gaps are introduced by spin-orbital coupling within the degenerate states of M$'$ and M$''$, while band-inversion occurs at the Γ-point among the degenerate and non-degenerate orbitals; driven by hybridization of the neighboring orbitals. Phonon-dispersion calculations show that the predicted topological insulators are structurally stable. Tungsten-based MXenes with large band-gaps were suggested to be promising for topological applications at room temperature. The electronic structures of thicker ordered compositions, $M'_2M''_2C_3O_2$, show that they are non-trivial topological semimetals. Both Mo_2TiC_2 and $Mo_2Ti_2C_3$ can be functionalized with a mixture of fluorine, oxygen and hydroxyl ions.

WC

Magnetoresistance, Hall effect and de Haas-van Alphen effect studies[204] of single crystals show that, when the magnetic field is rotated in the plane perpendicular to the current, the carbide exhibits a field induced metal-to-insulator like transition and a large non-saturating quadratic magnetoresistance at low temperatures. When the magnetic field is parallel to the current, a marked negative longitudinal magnetoresistance can be observed only for a certain direction of current flow. Hall-effect data indicate that the material is a perfect compensated semimetal. Analysis of the de Haas-van Alphen oscillations reveal that the material is a multi-band system having small cross-sectional areas of Fermi surface and light cyclotron effective masses. Triply degenerate nodal points near to the Fermi level have been observed by means of angle-resolved photo-emission spectroscopy. By solving the Boltzmann transport equation, using first-principles calculations, an investigation could be made[205] of the phonon transport properties. The predicted room-temperature lattice thermal conductivities along the a- and c-directions were 1140.64 and 1214.69W/mK, respectively. In spite of their similar crystal structures, the thermal conductivity of WC is more than two orders of magnitude greater than that of WN. Also unlike the case of WN, a large acoustic-optical gap prevents the acoustic+acoustic → optical scattering, which gives rise to very long phonon lifetimes and the ultra-high lattice thermal conductivity of WC. Again unlike the case of WC, the density of states of WN at the Fermi level becomes very sharp, leading to the destabilization of WN and producing soft phonon modes. Small G and C_{44} elastic constants limit the stability of WN as compared with WC.

Angle-resolved photo-emission spectroscopy has revealed triply degenerate nodal points near to the Fermi level, in which the low-energy quasi-particles can be described[206] as being three-component fermions which are distinct from Dirac and Weyl fermions. Topological surface states, the constant-energy contours of which constitute pairs of Fermi arcs connecting the surface projections of the triply degenerate nodal points, confirm the non-trivial topology of the new semimetal state.

First-principles calculations have shown[207] that the phonon spectra of the WC-type materials, TiS, ZrSe, HfTe, also exhibit unique triply-degenerate nodal points and single two-component Weyl-points of THz frequency. The quasi-particle excitations near to the triply-degenerate nodal points of phonons are three-component bosons which go beyond the usual Dirac, Weyl and doubly-Weyl phonons. The TiS and ZrSe have five pairs of type-I Weyl phonon, and a pair of type-II Weyl phonons. The HfTe has only four pairs of type-I Weyl phonons, with non-zero topological charges. There are topological protected surface arc-states, on the (10•0) crystal surfaces, which connect two Weyl points of opposite charge and exhibit modes which propagate in essentially one direction on the surface.

Y_2C

In electrides such as this[208], the electrons act like anions and the bands which they occupy lie close to the Fermi level because the anionic electrons are weakly bound to the lattice. This property is conducive to the occurrence of the band inversions which are required by topological phases. Thus, Y_2C and Sr_2Bi are nodal-line semimetals, Sc_2C is an insulator, HfBr is a quantum spin Hall system and LaBr is a quantum anomalous Hall insulator. First-principles calculations predict[209] the existence of a topological nodal line semimetal phase in two-dimensional compounds of the form, X_2Y, where X is calcium, strontium or barium and Y is arsenic, antimony or bismuth, in the absence of spin-orbital coupling and with a band inversion at the M-point. A non-trivial Z_2 invariant of unity remains, although a tiny gap appears at the nodal line when spin-orbital coupling is included. Mirror-symmetry and electrostatic interactions, which can be modified by straining, are deemed to be responsible for the non-trivial phase.

Chalcogenides

$Bi_4(Se,S)_3$

This compound can be regarded as being a 1:1 natural superlattice of alternating Bi_2 layers and Bi_2Se_3 layers, with the inclusion of sulfur permitting the growth of large crystals having the composition, $Bi_4Se_{2.6}S_{0.4}$, as revealed by spin- and angle-resolved photo-emission studies of this topological semimetal. Such crystals cleave through the interfaces between the Bi_2 and Bi_2Se_3 layers, with the resultant surfaces having alternating bismuth and selenium terminations[210]. The resulting terraces, observed by photo-emission electron microscopy, are suitable for the study of one-dimensional topological effects. The electronic structure, according to spin- and angle-resolved photo-emission spectroscopy, indicates the existence of a surface state which forms a large hexagonal Fermi surface around the Γ-point of the surface Brillouin zone; with the spin structure indicating that the material is a topological semimetal.

Cu_2S

A class of three-dimensional d-orbital topological materials has been identified in antifluorite crystals of this family. In Cu_2S, each copper atom is surrounded by a tetrahedron of sulfur atoms. The tetrahedral crystal field splits the copper 3d-orbitals into e_g and t_{2g} states; the latter having a higher energy. Thanks to the t_{2g} states, the interaction between the two copper atoms in the primitive cell leads to bonding and antibonding states, with the latter being energetically superior. The copper d-orbitals have a higher energy than that of sulfur 3p-orbitals, such that p-d hybridization pushes the t_{2g} antibonding states closer to the Fermi level. Following on from the properties of low-energy t_{2g} states, their phases are determined only by the sign of the spin-orbital coupling: giving a topological insulator in the case of negative spin-orbital coupling, and giving a topological semimetal for positive spin-orbital coupling. Both of them have Dirac-cone surface states, but differing helicities. In the presence of broken inversion symmetry, the semimetal has a nodal box which consists of butterfly-shaped nodal lines that are resistant to spin-orbital coupling. Further breaking of the tetrahedral symmetry by straining leads to an ideal Weyl semimetal having four pairs of Weyl points. The Fermi arcs co-exist with a surface Dirac cone on the (010) surface, as required by a Z_2 invariant[211].

HfSiS

The three-dimensional Fermi-surface topology of this material consists of asymmetrical electron and hole pockets, as indicated by a large anisotropy of the magnetoresistance[212]. A very low carrier effective mass and a π-Berry phase originate from a Dirac-type dispersion of the bands. The Berry-phase influence can be seen in the amplitudes of de Haas-van Alphen and Shubnikov-de Haas oscillations at up to 20K in a field of 7T. A linear dispersion near to the Fermi level results from an interplay between the orbitals of Si-square nets and hafnium atoms.

HgTeS

Electronic structure calculations have shown[213] that strained $HgTe_xS_{1-x}$ alloys have a surprisingly rich topological phase diagram (figure 21). In the strong topological insulator phase, the spin chirality of the topological non-trivial surface states can be reversed by varying the x-value and the strain. Positive or negative strain could be introduced by using a suitable substrate. Two semimetallic topological phases, a Dirac semimetal and a Weyl semimetal, were predicted to exist. In the case of Weyl semimetals, the surface contribution was expected to be very anisotropic because of the shape of the Fermi arcs, and this could be detected by making surface-sensitive measurements. On the other hand, the anisotropy should essentially vanish in the case of topological Dirac semimetals, since the Fermi lines are closed. In the case of strong topological insulator phases, the bulk contribution should vanish because of the insulating nature of the bulk. This would not be true of semimetallic phases. Assuming that the lattice constant of the present material varies linearly as a function of the x-value, it was suggested that GaSb, with its lattice constant of 6.1Å, could serve as a substrate that was likely to induce the Weyl semimetal phase. On the other hand, in order to induce the topological Dirac semimetal phase, $Cd_{0.7}Zn_{0.3}Te$ with its lattice constant of 6.364Å, was suggested to be a suitable substrate.

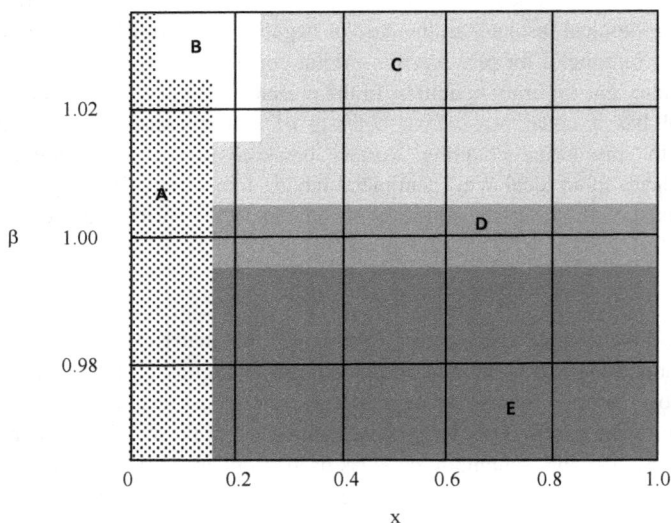

Figure 21. Phase diagram of strained $HgTe_xS_{1-x}$ compositions.
A: strong topological insulator ($c_m=+1$), B: Weyl semimetal,
C: strong topological insulator ($c_m=-1$), D: normal semimetal, E: Dirac semimetal

InNbS$_2$

First-principles density functional calculations show[214] that, in the absence of spin-orbital coupling, $InNbX_2$, where X is sulfur or selenium, exhibits nodal lines which are protected by mirror-symmetry. When spin-orbital coupling is accounted for, the Dirac rings in

InNbS$_2$ split into two Weyl rings. All other known nodal-line materials normally require an absence of spin-orbital coupling. However, spin-orbital coupling breaks the nodal lines in InNbSe$_2$ and the compound becomes a type-II Weyl semimetal having 12 Weyl points in the Brillouin zone.

PbTaS$_2$

First-principles calculations have shown[215] that this non-centrosymmetric material is a topological nodal line semimetal. In the absence of spin-orbital coupling, one band inversion occurs about a high-symmetry H-point; leading to the formation of a nodal line. The latter is stable, and is protected against gap-opening by mirror reflection symmetry, even in the presence of strong spin-orbital coupling. The material also involves complicated drum-head surface states, inside or outside of the projected nodal ring, depending upon the nature of the surface termination.

Pd$_3$Bi$_2$S$_2$

The transport properties of monocrystalline samples of this material were determined[216], as being one which has been predicted to contain an electronic phase that goes beyond the three-dimensional Dirac and Weyl semimetals. As in the case of some topological systems, its resistivity exhibits a field-induced metal-to-semiconductor cross-over at low temperatures. A high anisotropic non-saturating magnetoresistance has been detected in the transverse experimental configuration. At 2K, in a field of 9T, the magnetoresistance attains about 1100%. The Hall resistivity indicates the existence of two types of charge carrier. In spite of their high (10^{21}/cm^3) density, the mobility of the charge carriers was of the order of 7.5×10^3 cm^2/Vs for holes and 3×10^3 cm^2/Vs for electrons.

ZrSiS

It has been shown that the thermoelectric power (figure 22) is a very sensitive means for studying the quantum oscillations that are detectable in this material at temperatures as high as 100K. Two of these oscillations arise from three- and two-dimensional electronic bands, each of which exhibit linear dispersion and the additional Berry phase which is predicted to exist in materials possessing a non-trivial topology[217]. Since the properties of topological fermions are revealed by quantum oscillations, de Haas-van Alphen oscillation studies were made of this topological Dirac nodal-line semimetal. The angular dependence of the de Haas-van Alphen oscillations revealed[218] anisotropic Dirac bands in the material and the occurrence of very strong Zeeman splitting in low magnetic fields.

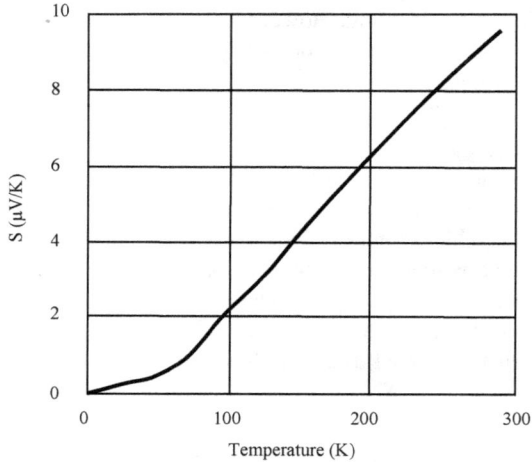

Figure 22. Temperature dependence of the a-axis thermoelectric power of ZrSiS in a zero magnetic field

The Landé g-factor, estimated from the separation of Zeeman splitting peaks, could be as large as 38. Analysis of the de Haas-van Alphen oscillations also revealed an essentially zero effective mass and a very high quantum mobility of the Dirac fermions. The topological properties of fermions arise from their low-energy Dirac-like band dispersion and associated chirality. Although initially limited to points, extension of the Dirac dispersion to lines and loops can occur. Further quantum-oscillation studies[219] of this nodal-loop semimetal in high magnetic fields revealed a marked enhancement of the effective mass of quasi-particles situated close to a nodal loop. At a threshold field, magnetic breakdown occurred across gaps in the loop structure, with orbits that enclosed various windings around its vertices; each winding generating an additional π Berry phase. The amplitudes of the breakdown orbits also exhibited an anomalous temperature dependence.

Ag$_2$Se

This topological semimetal has a Kramers-Weyl node at the origin in momentum space, plus quadruplet spinless Weyl nodes which are annihilated by spin-orbital coupling. Macrosized monocrystalline β-Ag$_2$Se exhibited[220] giant Shubnikov-de Haas oscillations in the longitudinal magnetoresistance behaviour, and these arose from a small electron pocket that can be driven beyond the quantum limit by a magnetic field of less than 9T. The pocket was a remnant of spin-orbital annihilated Weyl nodes and enclosed a Berry-phase structure. A negative longitudinal magnetoresistance was observed when the magnetic field was beyond the quantum limit.

Bi₄Se₃

Bi_4Se_3

Although three-dimensional topological insulators or semimetals which involve topological surface states are termed gap-less if time-reversal symmetry is maintained, this can be incorrect when surface-state degeneracies occur away from time-reversal invariant momenta. The mirror invariance of the system then becomes essential to the protecting of the existence of a superficial Fermi-surface. Such a situation exists in the case of the present strongly topological semimetal. Angle-resolved photo-emission spectroscopy and first-principles calculations[221] reveal a partial gapping of surface bands on the Bi_2Se_3 termination of $Bi_4Se_3(111)$. Here, an 85meV gap along $\overline{\Gamma K}$ closes down to zero towards the mirror-invariant $\overline{\Gamma M}$ azimuth. This is attributed to an inter-band spin-orbital interaction which mixes states of opposite spin-helicity.

$HgCr_2Se_4$

It was noted in the early days of the present field that, in three-dimensional momentum space, a topological phase boundary which separated Chern insulating layers from normal insulating layers could exist where the gap had to be closed and result in a so-called Chern-semimetal state having topologically unavoidable band-crossings at the Fermi level. This state would be a condensed-matter embodiment of Weyl fermions in (3+1) dimensions could be expected to exhibit features like magnetic monopoles and Fermi-arcs. At that time, it was predicted[222] that such a quantum state could exist in ferromagnetic compound $HgCr_2Se_4$, with a single pair of Weyl fermions separated in momentum space, and that the quantum Hall effect could occur without any external magnetic field. A complete classification of two-band k•p theories at band-crossing points in three-dimensional semimetals, possessing n-fold rotation symmetry and broken time-reversal symmetry, was also carried out some time ago[223]. This classification revealed the existence of new three-dimensional topological semimetals which were characterized by $C_{4,6}$-protected double-Weyl nodes with a quadratic in-plane (along $k_{x,y}$) dispersion, or C_6-protected triple-Weyl nodes with cubic in-plane dispersion. Application of this theory confirmed that the present three-dimensional ferromagnet is a double-Weyl metal, protected by C_4 symmetry. If the direction of the ferromagnetism is shifted from the [001]-axis to the [111]-axis, the double-Weyl node then splits into four single Weyl nodes; as dictated by the S_6 point group of that phase. It was also noted that quantum phase transitions were induced by applying strain: because the multi-Weyl nodes are protected by C_n invariance, an applied strain which breaks the symmetry can split a multi-Weyl node into several single Weyl nodes. This might be done by effectively adding a C_4 breaking term to the effective Hamiltonian around a double-Weyl node. Also, a double or triple Weyl node breaks into two or three Weyl nodes, respectively, under anisotropic straining in the xy-plane. This will generally cause a change in the density-of-states near to the node and can be expected to affect the bulk transport properties. There have been extensive studies of the magnetoresistance of this material but it seems that there has so far been little effort made to link the results to topological features. Extreme magnetoresistance in n-type samples has been reported[224], where low-density conduction electrons were exchange-coupled to a three-dimensional Heisenberg ferromagnet with a Curie temperature of 105K. Near to room temperature, the electron transport exhibited normal semiconducting behavior. Below the Curie temperature, the magnetic susceptibility deviated from the Curie-Weiss law, and the transport properties exhibited extreme magnetoresistance until a transition to metallic conduction occurred at lower temperatures. It was suggested that spin correlations were important, not only when near to the critical point but also over a wide temperature range of the paramagnetic phase. An analytical description was derived[225] for the magnetoresistance behaviour at low temperatures.

PbTaSe₂

The experimental observation of co-existing lead concave bands and of tantalum convex bands centred at the K-point in this non-centrosymmetric superconductor was in good agreement with first-principles band-structure predictions, and confirmed the occurrence of unusual ring-shaped topological nodal-line states; the nodal rings being protected by the reflection symmetry of the system[226]. The rings were also associated with drumhead-like surface states, rather in the way that edge states and nodal points are connected in graphene. In view of the one-dimensional nodal characteristics of the bulk band, and the two-dimensional topological drum-head surface states, topological nodal-line semimetals were here recognised as being a distinct class of topological material which was distinct from Weyl semimetals and topological insulators. The nodal-line states have an extra degree of freedom in the form of the finite size of the nodal line. Interaction-induced instabilities are also more likely to occur in nodal-line states, than in Weyl semimetals, because of the higher density of states at the Fermi energy. Superconduction can be induced moreover by intercalating lead layers into $TaSe_2$ and benefiting from the Pb-conducting orbitals in forming topological nodal-line states. In the case of intrinsic superconductivity, it is considered to be possible that helical superconductivity and p-wave Cooper pairing could exist here without the aid of the proximity effect.

ZrSiSe

A new type of surface state, the so-called floating band, has recently been discovered in this nodal-line semimetal, space group P4/nmm, and is expected to exist in many other non-symmorphic crystals. Scanning tunnelling microscopy has been used to measure the quasi-particle interference of the floating-band state on the (001) surface[227]. This also detected a rotational symmetry-breaking interference, a healing effect and a half-missing type of anomalous umklapp scattering. Simulations and theoretical analysis confirmed that these phenomena were characteristic properties of a floating-band surface state. The half-missing Umklapp process arises from a non-symmorphic effect which potentially exists in materials whose lattices contain glide mirrors, $(Mz|\frac{1}{2}\frac{1}{2}0)$. In nodal-line topological semimetals where band-touchings form nodal lines or rings, a loop which encloses a nodal line can permit an electron to accumulate a non-trivial π Berry-phase such that the phase-shift in Shubnikov-de Haas oscillations then reveals the presence of a nodal-line semimetal. On the other hand, experiment can indicate contradictory phase-shifts; especially in materials such as the present one or, indeed, in analogues produced by replacing the zirconium with hafnium, the silicon with germanium or the sulfur with selenium or tellurium. Systematic calculations have been made[228] of the Shubnikov-de Haas oscillation of resistivity, in a magnetic field normal to the nodal-line plane, for a generic model of a nodal-line semimetal. This furnished general rules for determining the phase-shifts in arbitrary situations and applying them to the ZrSiS and Cu_3PdN systems. Depending upon the magnetic-field direction, carrier type and cross-section of the Fermi surface, the phase-shift could be very different from that for normal electrons and Weyl fermions. Materials of the form, WHM, where W is zirconium, hafnium or lanthanum, H is silicon, germanium, tin or antimony and M is oxygen, sulfur, selenium or tellurium, are predicted to be topological materials. They offer accurate adjustment of the spin-orbital coupling, lattice constant and structural dimensionality and permit study of the influence of those parameters upon the topological semimetal state. High-field quantum-oscillation studies of ZrGeM, where M was sulfur, selenium or tellurium, have revealed[229] properties which are consistent with theoretically predicted topological semimetal states.

CeSbTe

A wide range of different topological semimetal states can exist in this material[230], and it can exhibit various types of magnetic order (figure 23) upon applying a small field. This allows the electronic structure to be driven through a number of topologically distinct phases, such as a non-symmorphic magnetic topological phase having an eight-fold band crossing at a high-symmetry point.

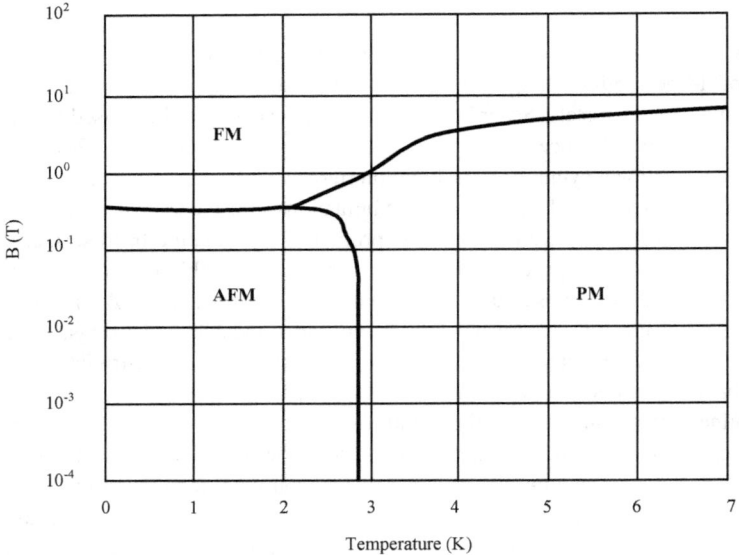

Figure 23. Magnetic phase diagram of CeSbTe
FM: ferromagnetic, AFM: antiferromagnetic, PM: paramagnetic

Fe₃GeTe₂

The van der Waals material, $Fe_{3-x}GeTe_2$, is likely to be a ferromagnetic nodal-line semimetal which exhibits a very large anomalous Hall effect[231]. The latter effect, driven by magnetic order, links spin–orbit coupling and quantum magnetism to differential geometry and topology. The present material is a three-dimensional itinerant ferromagnet which has very high figures-of-merit that are associated with the two-dimensional quantum anomalous Hall state. Among other factors, the intrinsic anomalous Hall effect is a measure of the Berry curvature of a band structure. The Berry curvature measures Berry-phase accumulation during a cyclic and adiabatic evolution through parameter space. In the case of electrons in crystals, the parameter space for electronic wave-functions is the momentum space and the Berry curvature governs the anomalous velocity of charge carriers. This velocity is transverse to the Berry-curvature vector and to the electric field which causes electron transport; thus producing the Hall effect. The anomalous Hall current is obtained when these velocities are summed over all electrons present in the crystal. A large Berry-curvature is associated with band degeneracies and is large in those regions of momentum space where two or more bands are close in energy. In the extreme case of band-touching at isolated points in momentum space, Berry curvature acts like a source or sink. The Berry curvature is singular near to such points and there resembles the magnetic field of a magnetic monopole (figure 24).

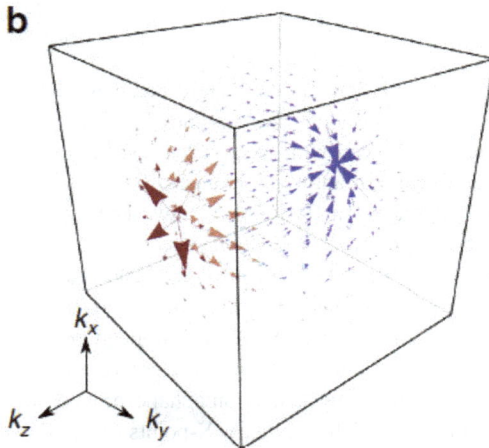

Figure 24. Vector plot of the Berry curvature in momentum space The arrows show that the flux of the Berry curvature flows from one (red) monopole to another (blue) in a topological semimetal. [reproduced under Creative Commons Licence from Li, H. et al. Negative magnetoresistance in Dirac semimetal Cd₃As₂. Nat. Commun. 7:10301 doi: 10.1038/ncomms10301 (2016).]

Theoretical considerations, magnetotransport and angle-resolved photo-emission spectroscopic data suggest[232] that this material is possibly a ferromagnetic nodal-line semimetal. The spin degree of freedom is entirely quenched by the large ferromagnetic polarization, while line degeneracy is protected by the crystalline symmetries which link two orbitals in adjacent layers. This orbital-driven nodal line is adjustable via the spin orientation that is due to spin–orbital coupling, and produces a large Berry curvature, which leads to a large anomalous Hall current, angle and factor. This indicates that ferromagnetic topological semimetals have potential for spin- and orbital-dependent electronic applications. Time- and angle-resolved photo-emission spectroscopy is the most direct means for probing the effects of optical excitation upon the band structure of a material[233]. The time-resolved technique, applied in the extreme ultra-violet range, accesses the ultra-fast dynamics over the entire Brillouin zone. It has also been used to measure the ultra-fast electronic response of ZrSiTe; a topological semimetal which is characterized by the appearance of linearly dispersing states, located at the Brillouin-zone boundary.

GdSbTe

Angle-resolved photo-emission spectroscopy has revealed[234] the existence of an antiferromagnetic topological nodal-line semimetallic state in this material. It has a non-symmorphic tetragonal unit cell with space group P4/nmm, in which Gd-Te bilayers are sandwiched between layers of antimony atoms to form a square net. At above 15K, the magnetic susceptibility obeys a Curie-Weiss law. The magnetic field variation of the magnetization exhibits a small inflection near to 1.5T which is attributed to a metamagnetic-like phase transition. In stronger fields, the material exhibits a slight upward behavior, indicating that a field-induced ferromagnetic arrangement of the gadolinium magnetic moments can be obtained in magnetic field strengths much greater than 5T. Detailed study of the electronic structure revealed the presence of multiple Fermi-surface pockets, including a diamond-shape around the zone center and small circular pockets around the high-symmetry X-points of the Brillouin zone. There was a Dirac-like state at the X-point of the Brillouin zone and an effect of magnetism along the nodal-line direction. There is a stable Dirac-like state below and above the magnetic transition temperature of 13K.

MoTe$_2$

The low-temperature orthorhombic structure of this material, when studied by using X-ray diffraction at 100K, is calculated[235] to contain four type-II Weyl points between the nth and (n±1)th bands, where n is the total number of valence electrons per unit cell. Other Weyl points and nodal lines between various other bands also appear close to the Fermi level due to a complicated topological band structure. Strain-driven topological phase transitions offer the possibility of creating various topological semimetal phases. In the absence of strain, the number of surface Fermi arcs is two; the smallest number of arcs which is consistent with time-reversal symmetry. It has been noted that superconductivity at 2.1K occurs in Te-deficient orthorhombic $MoTe_{2-x}$ with intrinsic electron-doping, whereas stoichiometric monoclinic $MoTe_2$ exhibits no superconducting state down to 10mK but has a magnetoresistance of 32000% at 2K in a magnetic field of 14T; the latter originating from almost-perfect compensation of the electron and hole carriers. Scanning tunnelling spectroscopy, synchrotron X-ray diffraction and theoretical calculations have shown[236] that tellurium vacancies trigger the superconductivity via intrinsic electron-doping and structural phase changes below 200K. This tellurium-vacancy induced superconductivity is attributed to the fact that the orthorhombic phase has been predicted to be a Weyl semimetal under electron-doping. Such chalcogen-induced superconductivity thus provides a potential route to combining superconducting and Weyl semimetal states in two-dimensional van der Waals materials.

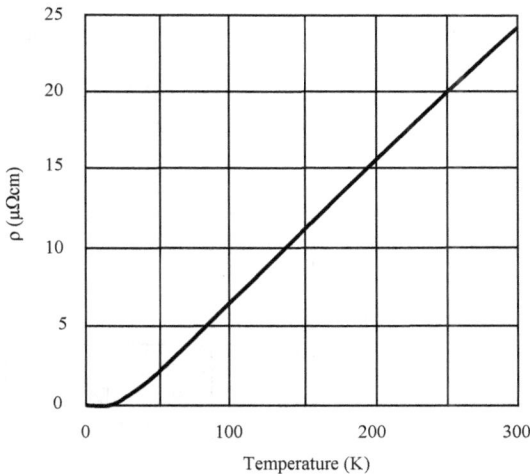

Figure 25. Resistivity of $PtTe_2$ (zero field) as a function of temperature

$NaCu_3Te_2$

An effective approach has been reported[237] for searching for ideal triply-degenerate-point semimetals via selective band-crossing between antibonding s-orbitals and bonding p-orbitals along a line in momentum space having C_{3v} symmetry. By using this approach, the present material and its relatives have been identified as being ideal triply-degenerate point semimetals, in which just two pairs of triply degenerate points are located around the Fermi level. A fundamental mechanism is found to modulate the energy-splitting between a pair of triply-degenerate points.

PtTe$_2$

Experimental evidence has been found for the existence of type-II Dirac fermions in bulk stoichiometric PtTe$_2$ monocrystals[238]. Angle-resolved photo-emission spectroscopic measurements and first-principles calculations indeed revealed the presence of a pair of strongly tilted Dirac cones along the Γ-A direction, confirming that this material is a type-II Dirac semimetal. In addition, the electrical resistivity (figure 25) and Hall effect were investigated in high-quality monocrystalline specimens. The magnetoresistance (figure 26) was found to be over 3000% at 1.8K in a field of 9T, and was unsaturated in strong fields over the entire temperature range which was investigated[239]. The magnetoresistance isotherms exhibited Kohler-type scaling with an exponent of 1.69. In an applied magnetic field, the resistivity contained the low-temperature plateau which is a characteristic of a topological semimetal. In strong fields, well-resolved Shubnikov-de Haas oscillations having two principle frequencies were found.

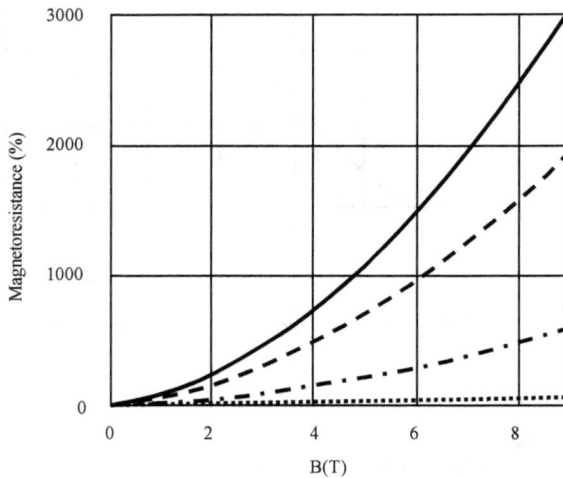

Figure 26. Magnetoresistance of PtTe$_2$ as a function of magnetic field strength Solid line: 1.8K, dashed line: 15K, dash-dot line: 26K, dotted line: 50K

Materials Research Forum LLC
doi: http://dx.doi.org/10.21741/9781644900154

Analysis of these yielded charge mobilities which were of the order of $1000cm^2/Vs$ and effective charge-carrier masses of 0.11 and $0.21m_e$. On the other hand, the Berry phases which were deduced indicated a trivial nature of the electronic bands which were involved in the Shubnikov-de Haas oscillations. Finally, the Hall-effect results confirmed the multi-band nature of the electrical conductivity, with moderate charge compensation. Noting that previous studies were limited mainly to those conducted on bulk crystals and exfoliated flakes, ultra-thin two-dimensional nanosheets of controllable thickness were used in the investigation of thickness-dependent electronic properties. Nanosheets were grown, using chemical vapor deposition, in an hexagonal or triangular geometry having a lateral dimension of up to 80μm. The thickness could meanwhile be chosen to lie between 20 and 1.8nm by reducing the growth temperature or by increasing the flow rate of the carrier gas. The resultant two-dimensional nanosheets were high-quality monocrystals. Raman spectroscopy indicated characteristic E_g and A_{1g} vibrational modes at about 109 and 155/cm, respectively[240]. A systematic red-shift occurred with increasing thickness. The two-dimensional nanosheets had a conductivity of up to 2.5 x 10^6S/m (figure 27). The conductivity and its temperature-dependence could be varied widely by changing the thickness. The nanosheets also had a breakdown current density of up to 5.7 x 10^7A/cm^2; currently the highest known for a two-dimensional metallic transition-metal dichalcogenide.

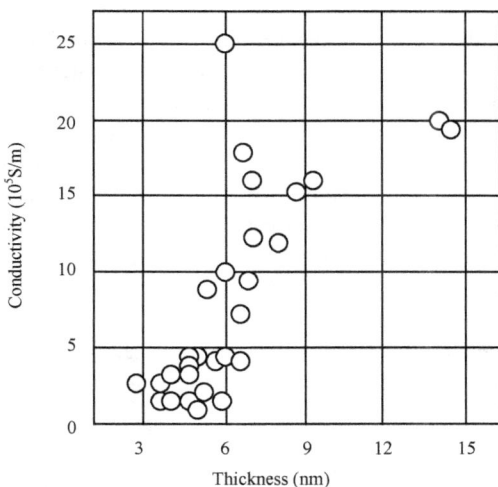

Figure 27. Conductivity of PtTe$_2$ as a function of nanofilm thickness

PtSeTe

The paths which contain type-II Dirac points in the $PtSe_2$ group of materials also exhibit spatial symmetry. Due to Kramers degeneracy, resulting from the combination of inversion symmetry with time-reversal symmetry, the crossing-points within are of Dirac-type. First-principles calculations predict[241] that $PtSe_2$-group materials should undergo a topological transition if the inversion symmetry is broken. Thus the Dirac fermions in this group of materials split into triply degenerate points in PtSeTe-group materials such as PtSSe, PtSeTe and PdSeTe. This is a useful example of Dirac points transforming into triply degenerate points.

SrTe

The electronic properties and Fermi-surface topology under pressures of up to 50GPa have been predicted[242] on the basis of density functional theory. The material is expected to undergo a structural phase transition from an NaCl-type to a CsCl-type structure at 14.7GPa. The ambient and high-pressure phases were both indirect band-gap semiconductors. During further compression, the CsCl-structured phase became a non-trivial topological semimetal.

$TaTe_4$

In its complicated structure, a linear chain of tantalum atoms is surrounded by four Te_2 dimer chains with the distance of the Te_2 dimer being smaller than twice the radius of a tellurium atom. The overall structure is then made of a parallel arrangement of the Ta-atom linear chains, surrounded by their Te_2 dimers. It is commensurately modulated at room temperature and changes into another commensurate structure upon heating. Powder X-ray diffraction reflections indicate a tetragonal space group, P_{4cc}, with lattice parameters of a = b = 6.515Å and c = 6.815Å. The resistivity and magnetoresistance of single crystals have been measured in material having a charge density wave transition temperature of 475K[243]. A magnetoresistance of about 1200% in a magnetic field of 16T was measured at 2K (figure 28). A field-induced universal topological-insulator resistivity, with a plateau at about 10K and a high carrier quantum mobility in the plateau region were present. The quantum oscillations reflected the angle-dependence of a two-dimensional Fermi surface.

$TaIrTe_4$

Density functional theory calculations within the local density approximation, together with scalar and full four-component relativistic methods, showed[244] that this compound is a type-II Weyl semimetal having the simplest possible arrangement, of only four well-separated Weyl points, under the symmetry constraints of the compound. Topological Fermi arcs were present on both surfaces of a naturally cleaved ideal semi-infinite slab, as well as for other surface terminations. The length of the emergent Fermi-arc of the (001) surface amounted to about 1/3 of the Brillouin-zone extent in the b-direction. The energy-range within which the Fermi-arcs were expected to be detectable ranged from 50 to 82meV above the Fermi level. It was suggested that the existence of the above particular Weyl points is a general characteristic of this family of compounds. It was expected that slight modification of the band-filling, such as that caused by doping, could bring the bulk Weyl points - and the surface arcs which connected them – to, or below, the Fermi level ... where spectroscopy could then resolve them.

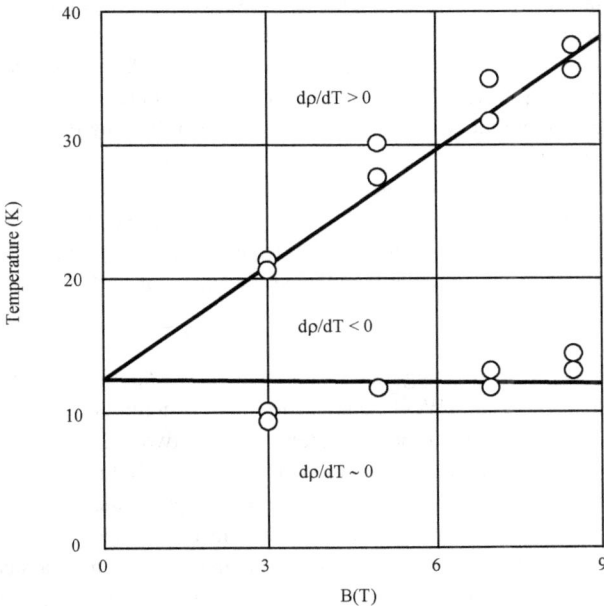

Figure 28. Phase diagram of TaTe₄ single crystals

WTe_2

This material was originally highlighted[245] as an example of a topological semimetal which featured a new particle as a low-energy excitation around a type-II Weyl point.

The existence of the latter points causes many of its physical properties to be very different to those of the usual Weyl semimetals having point-like Fermi surfaces. Proposed here was the existence of an overlooked type of Weyl fermion that appeared at the boundary between electron and hole pockets in a new phase of matter. It was noted that this particle was overlooked because it did not have an analogue in high-energy physics. By generalizing the Dirac equation, a new type of Weyl fermion could be identified. Materials which featured Weyl fermions had previously been thought to comprise Weyl points having a point-like Fermi surface (type-I), but it was realised that a type-II Weyl point, still a protected crossing, appeared at the contact of electron and hole pockets in type-II Weyl semimetals. Strong proximity-induced superconductivity has been detected[246] in this type-II Weyl semimetal by using a van der Waals hybrid structure which was obtained by mechanically placing $NbSe_2$ onto various thicknesses of WTe_2. When the latter thickness reached 21nm, a superconducting transition occurred around the critical temperature of $NbSe_2$ with a gap amplitude of 0.38meV and an ultra-long proximity length of up to 7μm. Thicker (42nm) WTe_2 layers led to a critical temperature of 1.2K, a gap amplitude of 0.07meV and a proximity length of less than 1μm. Calculations which were based upon Bogoliubov-de Gennes equations predicted that the induced superconducting gap was an appreciable fraction of the $NbSe_2$ superconducting gap when the WTe_2 was thinner than 30nm. It then decreased quickly with increasing thickness, in qualitative agreement with experiment. A phase diagram of the magnetoresistance behaviour can be found under the entry for LaBi.

ZrTe

First-principles calculations suggested[247] that this is a topological semimetal which comprises six pairs of chiral Weyl nodes in its first Brillouin zone, and is different to other topological semimetals in that it contains another two pairs of massless fermions with triply degenerate nodal points. Mirror symmetry, three-fold rotational symmetry and time-reversal symmetry all required the Weyl nodes to have the same velocity vectors and to be located at the same energy level. The elastic and transport properties have been investigated by combining first-principles calculations with semi-classical Boltzmann transport theory[248]. The calculated elastic constants (table 5) proved its mechanical stability. It is found that spin-orbital coupling has a slight enhancement effect upon the Seebeck coefficient. Along the a(b) and c directions of pristine ZrTe, at 300K, it is equal to 46.26 and 80.20μV/K, respectively. Comparison of the experimental 300K electrical conductivity with the calculated value indicates a scattering time of 1.59×10^{-14}s. The predicted room-temperature electronic thermal conductivity along the a(b)- and c-directions is 2.37 and 2.90W/mK, respectively. The room-temperature lattice thermal conductivity is predicted to be 17.56 and 43.08W/mK along the a(b)- and c-directions, respectively; indicating very strong anisotropy. Isotope scattering produces an observable effect upon the lattice thermal conductivity. The average room-temperature lattice thermal conductivity is slightly higher than that of the isostructural semimetal, MoP. This is attributed to longer phonon lifetimes and smaller Grüneisen parameters.

Table 5. Elastic constants of ZrTe

C_{11}	140.82GPa
C_{12}	58.78GPa
C_{13}	88.81GPa
C_{33}	201.11GPa
C_{44}	110.36GPa
B	102.50GPa
G	61.26GPa
E_{xx}	97.83GPa
E_{zz}	122.07GPa
$v_{xy/yx}$	0.19
$v_{xz/yz}$	0.36
$v_{zx/zy}$	0.45

ZrTe$_5$

It is found that the resistivity peak temperature of this highly anisotropic three-dimensional Dirac semimetal can be extensively varied by choosing the nanosheet thickness[249]. When the thickness was reduced from 160 to 40nm, the resistivity peak temperature steadily decreased from 145 to 100K. On the other hand, when the thickness was reduced to 10nm the peak temperature moved back up to a higher temperature. The system changed, from a topological semimetal having two types of carrier to a single band having the usual hole carriers, when the thickness was less than 40nm. Meanwhile the Fermi-level shifted continually downwards, from the conduction band to the valence band, with decreasing thickness.

ZrSnTe

High-field de Haas-van Alphen quantum-oscillation studies[250] have provided evidence for the existence of topological non-trivial bands in this material. The angular dependence of the quantum-oscillation frequency reveals a three-dimensional Fermi surface topology of this layered material which is due to strong interlayer coupling.

Metals and Alloys

CaPd

This relatively little investigated metal phase lies on the border between type-I and type-II nodal line topologies, where types are assigned according to the degree of tilt of the fermion cone. This critical-type nodal line material exhibits an unique non-trivial band-crossing which comprises a flat band and a dispersive band and leads to an unfamiliar fermionic state. It is currently proposed[251] that this intermetallic can be regarded as an existing type of topological metal with regard to the fermionic state; characterized by a critical-type nodal line in the bulk and a drum-head band structure at the surface.

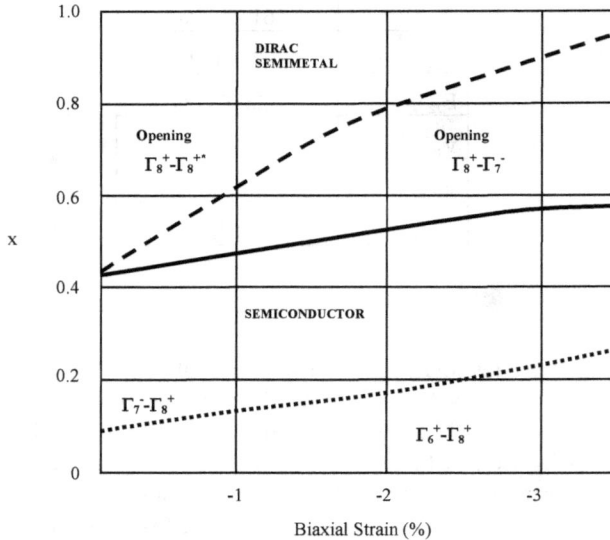

Figure 29. Phase diagram of $Ge_{1-x}Sn_x$

Co_2MnGa

Recalling that the band-crossings of a Dirac, Weyl or other fermion semimetal are zero-dimensional, and that the band-crossings of a nodal-line semimetal are one-dimensional closed loops, it has been suggested that the presence of perpendicular crystalline mirror planes can protect three-dimensional band-crossings which are characterized by the existence of non-trivial links such as a Hopf link or a coupled chain. This would create a range of novel types of topological semimetal. A non-trivial winding-number protects topological surface states which are distinct from those in topological semimetals that exhibit a vanishing spin-orbital interaction. Such non-trivial links can be created by adjusting the mirror eigenvalues which are associated with the perpendicular mirror planes. First-principles band-structure calculations predict[252] that the present ferromagnetic full-Heusler alloy would be a suitable material, and Hopf-link and chain-like bulk band-crossings and novel topological surface states have been identified.

$GeSn$

Theoretical studies of the electronic structures of $Ge_{1-x}Sn_x$ alloys ($0 \leq x \leq 1$), using the non-local empirical pseudopotential method, have shown[253] that relaxed material with x greater than 0.41 is a gap-less topological semimetal with band-inversion at the Γ-point. There is an indirect-to-direct band-gap transition at x = 0.085. In the case of strained samples, deposited onto a germanium substrate, the material is a semimetal with a negative indirect band-gap when x is greater than 0.43. Such strained material is always an indirect band-gap semiconductor when x is less than 0.43 (figure 29). The application of suitable biaxial compressive strains produces a topological Dirac semimetal which exhibits band inversion at the Γ-point and a pair of Dirac cones along the [001] direction.

Pt_3Sn

Recalling that topological insulators exhibit a topologically non-trivial band inversion, and that topological Dirac/Weyl semimetals exhibit so-called relativistic linear band-crossings, this material is interesting in that it exhibits both of the latter topological features simultaneously. First-principles calculations show[254] that it is a three-dimensional weak topological semimetal, with topologically non-trivial band inversion between the valence and conduction bands. The band structure also has type-II Dirac points at the boundary of two electron pockets. The lowest conduction bands develop type-II Dirac points along high-symmetry paths in the Brillouin zone, and the valence bands have a non-trivial band topology. The formation of the Dirac points can be understood in terms of the representations of relevant symmetry groups and compatibility relations. The topological surface states appear in accordance with the non-trivial bulk band topology.

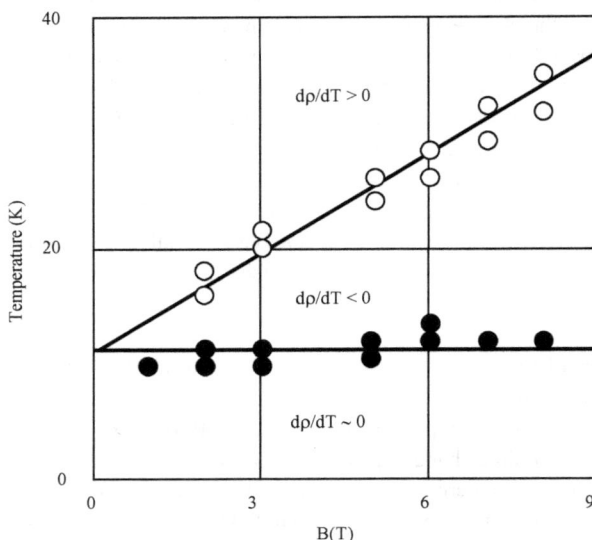

Figure 30. Temperature-field phase diagram of PtSn₄ monocrystals

PtSn₄

This material has an extremely large magnetoresistance and exhibits many topological features, such as Dirac node arcs in which the nodes form closed loops in momentum space. Angle-dependent magnetoresistance, Hall effect, heat-capacity measurements and first-principles calculations[255] have revealed anomalies at around 55K. The resistivity could be summarised by a phase diagram (figure 30). Marked changes in the transport results and heat capacity reflect successive Fermi surface reconstruction; implying the occurrence of a temperature-induced Lifshitz transition. Perfect compensation between electrons and holes has been found at around 30K. Effects are suggested to originate from a combination of electron-hole compensation and the orbital texture of the electron pocket.

RRh₆Ge₄

RRh_6Ge_4

Triply-degenerate fermions have interesting physical properties which are represented by triply-degenerate nodal points in topological semimetals. Space-group theory analysis offers a means of identifying topological semimetals with triply-degenerate nodal points that are located on a symmetry axis; a means which is applicable to both symmorphic and non-symmorphic materials. This, together with first-principles electronic structure calculations, predicts[256] that the present materials represent a class of triply-degenerate topological semimetals, where R is yttrium, lanthanum or lutetium, and the triply-degenerate nodal points are located on the Γ-A axis, not far from the Fermi level. Here, $LaRh_6Ge_4$ in particular has a pair of triply-degenerate nodal points which are located only about 3meV below the Fermi level.

Sn

High-resolution angle-resolved photo-emission study of the topological semimetal, α-tin, deposited onto InSb(001) permits observation of the topological surface state that is degenerate with the bulk band-structure and shows[257] that the former is unaffected by various surface reconstructions. The topological surface state has an almost ideal Dirac cone shape, with a circular Fermi contour. An unintentional p-type doping of the as-grown films was compensated by depositing potassium or tellurium post-growth, thus moving the Dirac point of the surface state to below the Fermi level. A two-dimensional feature appeared only away from the Γ-X direction in k-space. It is to be noted that, in Weyl, Dirac and nodal-line semimetals, the band-gap closes at points, or along lines, in k-space which are not necessarily located at high-symmetry positions in the Brillouin zone and it is thus not straightforward to find topological semimetals by first-principles calculation because the band structure tends to be calculated only along high-symmetry lines. The tellurium and potassium acted as electron-donors, and could shift the Dirac point by at least 50 and 215meV, respectively. Although it had the potential ability to break time-reversal symmetry, up to 0.25 of a monolayer of ferromagnetic iron impurity produced a slight n-type doping but did not have any greater effect upon the surface state above that of the potassium or tellurium. Measurement of the spin-momentum locking of electrons from the topological surface state, by means of spin-resolved photo-emission techniques, showed that the spin vector lies entirely in the plane but also has a finite radial component. It has a finite component that is antiparallel to the momentum. Although the topological surface state was obviously spin-polarized, the circular dichroism of the angular distribution vanished. Analysis of the decay of photoholes that were introduced during photo-emission studies helped to understand the many-body interactions of the system. The quasi-particle lifetimes which were deduced were similar to those in topological materials where the topological surface state is located within a bulk band-gap. The main decay process of the photoholes was attributed to intraband scattering, while scattering into bulk states was assumed to be suppressed due to the differing orbital symmetries of the bulk and surface states. In this connection, it has been demonstrated[258] theoretically, by analysing the edge states for an isolated node in a two-dimensional semimetal which is protected by chiral symmetry and characterized by a topological winding number, N, that the chiral structure of the nodes of nodal semimetals governs the existence and local properties of edge states near to the nodes. Consideration of the asymptotic chiral-symmetrical boundary conditions then showed that there are N+1 universal classes, with the class determining the numbers of flat-band edge states on each side of the node in the one-dimensional spectrum, and N giving the total number of edge states. The edge states of chiral nodal semimetals persist moreover within a finite stability region of chiral-asymmetrical parameters. It was shown that the Luttinger model, with a quadratic node for $j = 32$ electrons, effectively represents a three-dimensional topological semimetal and it was predicted that α-Sn, and many other semimetals which are described by it, are topological materials with surface states.

SrHgPb

This non-centrosymmetric hexagonal ABC-type material is proposed to represent yet another new type of topological semimetal which contains both Dirac and Weyl points in its momentum space. The material has a so-called stuffed-wurtzite lattice with an hexagonal structure in which the point group, $P6_3mc$ is isomorphic with C_{6v} and contains elements which involve a half-translation along the z-direction, such as the six-fold screw rotation, S_{6z}. The proper, C_{3v}, sub-group which does not involve fractional translations, is generated by a three-fold rotation, C_{3z}, and a mirror reflection, M_{yz}. The unit cell consists of two buckled HgPb layers and two strontium atoms which occupy the interstitial sites of the HgPb wurtzite lattice. The HgPb layer-buckling amounts to 0.78Å. The existence of symmetry-protected Dirac points is attributed to band inversion. They are located on the six-fold rotation z-axis, while six pairs of Weyl points - related by six-fold symmetry - are located on the perpendicular, $k_z = 0$, plane. Study of the electronic structure as a function of HgPb layer buckling, the origin of inversion symmetry-breaking, shows[259] that the coexistence of Dirac and Weyl fermions defines a phase which separates two, topologically distinct, Dirac semimetals. These are distinguished by the Z_2 index of the $k_z = 0$ plane and by the corresponding presence/absence of two-dimensional Dirac fermions on the side surfaces. In the limit of vanishing buckling, inversion symmetry is restored, and the space group becomes $P6_3/mmc$. The evolution of the total energy as a function of buckling confirms that the buckled phase is lower in energy than is the centrosymmetric phase. At some stage, a topological transition occurs at which Weyl points are created.

Yb

Resistivity measurements under pressure have been made[260] in the vicinity of the quantum critical point of face-centred cubic ytterbium. A band-gap opened up at about 12kbar, and this opening was not directly associated with a Lifshitz transition because transitions of pocket-vanishing type appeared before and after the gap opening; at about 10 and 14.3kbar. The metal-insulator transition that occurred at 14.3kbar could be perfectly scaled by using an approach derived for Wilson transitions. This was thought to be first observation of a metal-insulator transition with a scaling that entirely neglected electron-electron interactions. An anomaly was observed in the screening coefficient of the T^2 term in the resistivity of the metallic phase at low temperatures. The results were interpreted in terms of the effect of calcium impurity-doping. It was concluded that the T^2 resistivity behavior, which had previously been attributed to Baber scattering at low temperatures, might in fact originate from inelastic scattering on impurity phonons (Taylor-Koshino scattering).

Nitrides

Cu_3N

A class of Z_2 class of topological semimetals exhibiting vanishing spin-orbital interaction has been proposed[261]. These materials are characterized by the presence of bulk one-dimensional Dirac line nodes, and of two-dimensional nearly-flat surface states protected by inversion and time-reversal symmetries. The Z_2 invariants governed the presence of Dirac line nodes via the parity eigenvalues at parity-invariant points in reciprocal space. First-principles calculations predicted that Dirac line nodes would occur in Cu_3N, near to the Fermi energy, upon doping with non-magnetic transition metal atoms such as zinc and palladium. Topological semimetal surface states would appear in the projected interior of Dirac line nodes.

Cu_3PdN

This cubic (Pm$\overline{3}$m) anti-perovskite, has a nitrogen atom at the cube center, surrounded by octahedral copper atoms and with palladium at the cube corner. First-principles calculations show that the low-energy states are dominated by palladium 4d and palladium 5p orbitals and that band inversion occurs at the R-point. Without spin–orbital coupling, the occupied and unoccupied low-energy bands are triply-degenerate at the R-point. The band inversion is governed by the hopping parameters of p and d orbitals on palladium atoms. The Pauli matrices here characterize two bands arising mainly from the p_z and d_{xy} orbitals of palladium atoms. Closed nodal lines exist, as coexistence of time-reversal and spatial-inversion symmetries occurs in the case of band inversion and the absence of spin–orbital coupling. The surface states on the (001)-surface can be calculated by using a tight-binding Hamiltonian and maximum localized Wannier functions. Surface flat bands exist within the projected nodal-line ring. First-principles calculations indicate that, in the presence of spin–orbital coupling, a gap of about 0.062eV opens up in the R-X direction, where R and X are corner and face-center points, respectively, of the (001)-projected Brillouin-zone surface. The nodal point in the R-M direction is unaffected by spin–orbital coupling, where M is the mid-side point of the above Brillouin-zone surface. The nodal lines thus evolve into three pairs of Dirac points, and the (001)-direction surface band-structure has a gapped bulk state plus a surface Dirac cone due to topologically non-trivial Z_2 indices. These bulk Dirac cones are obscured by other bulk states and it is thus difficult to trace the detailed connections of Fermi-arcs in the Fermi surface plot, although amusingly-named eyebrow Fermi-arcs are discernible around the projected Dirac nodes.

GdN

In this compound, a half-metallic ferromagnetic material, splitting of triple nodal points arises from a spin-orbital coupling in which the size of the splitting ranges from 15 to 150meV; depending upon the magnetization orientation[262]. This permits a transition between a Weyl-point phase and a so-called nearly triple-nodal point phase which exhibits very similar surface spectra and transport properties to those of a true triple-node system. First-principles calculations show that emergent triple nodal points, in the absence of spin-orbital coupling, split into conventional Weyl points upon taking weak spin-orbital coupling into account. Some crossing-points on the C_4 rotation axis orthogonal to the magnetization direction remain in the nearly degenerate triple nodal point state. One essential requirement for the occurrence of certain band-crossings is the presence of a band inversion at the X-points which is sensitive to the magnitude of the lattice constant and the exchange-coupling strength (figure 31).

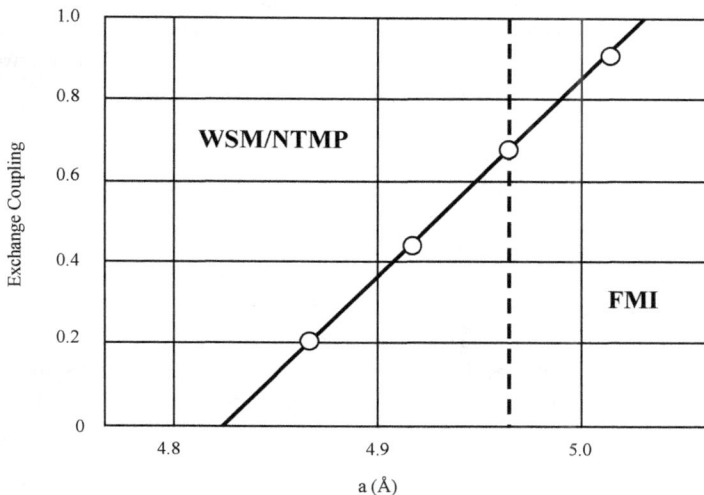

Figure 31. Phase diagram of GdN with the magnetic moment along [001]
Points indicate closure of the direct band-gap without spin-orbit coupling,
WSM: Weyl semimetal, NTNP: nearly triple nodal point,
FMI: ferromagnetic insulator, dashed line: equilibrium lattice constant

Li_2CrN_2

First-principles calculations suggest[263] that an alkali metal, deposited onto the surface of hexagonal (P6/m) XN_2 (X = Cr, Mo, W) nanosheets, can create topologically non-trivial phases. When spin-orbital coupling is ignored, the electron-like conduction band arising from nitrogen p_z orbitals can be considered to cross the hole-like valence band arising from X d_z^2 orbitals and thus create a topological nodal-line state in lithium-functionalized XN_2 sheets. The band crossing is here protected by the existence of mirror reflection and time reversal symmetries. The bands cross exactly at the Fermi level, and the linear dispersion regions of such band crossings can be as much as 0.9eV above the crossing. In the case of Li_2CrN_2, the results reveal the appearance of a Dirac cone at the Fermi level. Calculations also show that lattice compression decreases the thickness of a Li_2CrN_2 nanosheet and provokes a phase transition into a nodal-line semimetal. Changes in the band-gap of Li_2XN_2 at the Γ-point indicate that the non-trivial topological nature of

Li_2XN_2 nanolayers is stable over a wide strain-range. When spin-orbital coupling is included, the band crossing-point is gapped-out and this produces quantum spin Hall states in Li_2CrN_2 nanosheets, with an energy gap of 17.5meV. In Li_2MoN_2 however, the spin-orbital induced gap at the crossing-points is negligible.

MgTa$_2$N$_3$

Recalling that Dirac, triple-point and Weyl fermions are the three topological semimetal phases, as characterized by a decreasing degree of band degeneracy, this material is interesting in that all three types can be expected to occur in the single system, MgTa$_2$-$_x$Nb$_x$N$_3$. On the basis of symmetry analysis and first-principles calculations, the phase diagram of topological order can be mapped out[264] – as a function of alloy concentration and crystal symmetry (figure 32) – such that the intrinsic MgTa$_2$N$_3$ composition of highest symmetry exhibits the Dirac semimetal phase. This then transforms into the triple-point and Weyl semimetal phases with increasing niobium additions, which lowers the crystal symmetry.

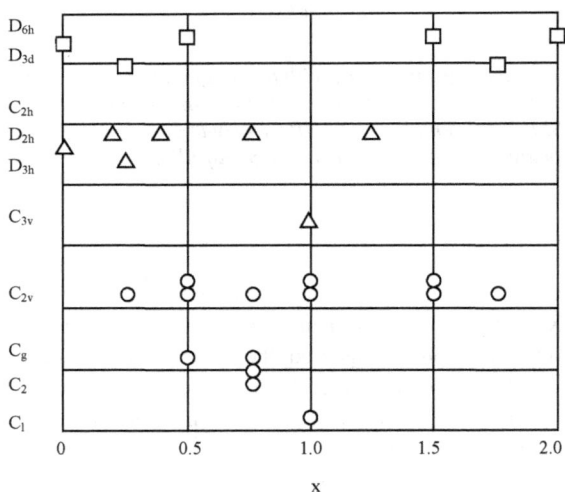

Figure 32. Symmetry-breaking effect of niobium-doping of MgTa$_{2-x}$Nb$_x$N$_3$
Squares: Dirac, Triangles: Triple, Circles: Weyl
Symmetry increases from bottom to top

Table 6. Lattice constants of TaN

Constant	Value (GPa)
G	233.16
B	260.26
C_{11}	566.40
C_{12}	128.23
C_{13}	62.41
C_{33}	706.39
C_{44}	215.02
$C_{6}6$	219.09
E_{xx}	534.04
E_{zz}	695.17

TaN

First-principles calculations and symmetry-analysis predicted[265] this material to be a topological semimetal which exhibited a then-new type of point node: the triply-degenerate nodal point. Each node is a band-crossing between degenerate and non-degenerate bands along the high-symmetry line in the Brillouin zone, and is protected by crystalline symmetries. Such nodes would always generate singular touching points between differing Fermi surfaces and a surrounding three-dimensional spin texture. Breaking of the crystalline symmetry by an external magnetic field, or by straining, produces various topological phases. Study of the Landau levels under low field conditions, along the c-axis, revealed a new quantum feature that was termed helical anomaly. The elastic properties and thermal conductivity have been investigated[266] by using first-principles calculations, a linearized phonon Boltzmann equation and a single-mode relaxation-time approximation. As expected, the calculated bulk modulus, shear modulus and C_{44} constant (table 6) indicated that the nitride is an essentially incompressible hard material. The room-temperature lattice thermal conductivity was predicted to be 838.62W/mK along the a-axis and 1080.40W/mK along the c-axis, and was thus very anisotropic. It was also several orders of magnitude higher than that of topological semimetals such as TaAs, MoP and ZrTe. The difference was attributed to the much longer phonon lifetimes in TaN. The long phonon lifetimes in turn result from the very different atomic masses of the tantalum and nitrogen atoms: this leads to a very large acoustical-optical band-gap and thus prohibits scattering between acoustic and optical phonon modes. It is predicted that isotope scattering will have little effect upon the lattice thermal conductivity, and that phonons with mean free paths that are greater than 20μm along the a-axis and greater than 80μm along the c-direction at 300K make little contribution to the total lattice thermal conductivity. A mass-difference factor (figure 33) can be defined, and is calculated by subtracting the lower mass from the higher one, and dividing by the lower mass.

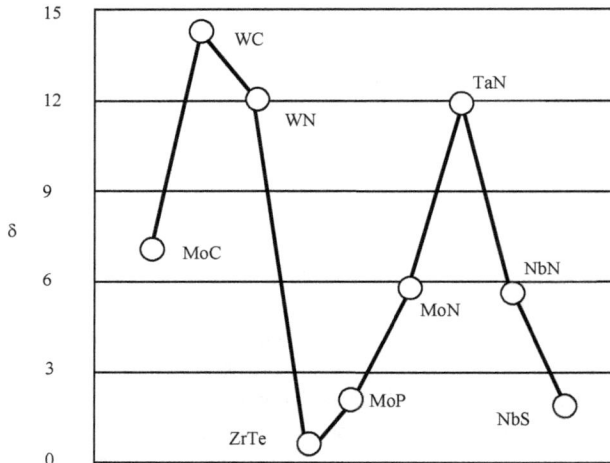

Figure 33. Mass-difference factor of WC-type topological metals

YN

First-principles calculations and model analysis have demonstrated[267] that strained YN can have an associated topological semimetal state, with an ultra-flat nodal ring, in the absence of spin-orbital coupling. Unlike the case of other nodal-line semimetals, the nodal ring here lies in a single plane of the Brillouin zone, with an energy variation of less than 0.3meV: the nodal ring is flat in both momentum space and energy space. By applying a uniaxial strain along the c-axis, it is possible to adjust the size of the nodal ring (figure 34) and potentially provoke a topological phase transition from nodal-ring semimetal to normal insulator. This nodal ring is topologically non-trivial, according to direct calculation of the topological invariant and the topological drum-head surface state.

Oxides

Various topological phases have been found in oxides, many of which involve heavy elements that exhibit strong spin–orbit coupling, such as iridium, osmium or bismuth. The electron correlations are relatively weak as compared with light transition elements, but their strength as compared with spin–orbit interactions is very important in creating novel electronic phases.

In the case of bilayer films of perovskite oxides such as $LaAuO_3$, topologically non-trivial band-gaps of the order of 50 to 300meV are expected and would be large enough to enable the quantum spin Hall effect at high temperatures. In these heterostructures, it may be possible to modify the topological electronic phase by applying an epitaxial strain or external gate voltage.

In a ZnO/(Mg,Zn)O hetero-interface, grown by means of molecular beam epitaxy, the electron mobility can exceed $1000000cm^2$/Vs. In ZnO, the high degree of control of the energy ratio between Zeeman and Landau splitting is good for studying the so-called non-abelian Pfaffian state. Topological electronic states can be described by the Dirac Hamiltonian, together with an additional one in three dimensions. If the extra Hamiltonian is an element of an abelian group, the electronic states become topologically non-trivial even in the absence of time-reversal and particle-hole symmetries. Symmetry-breaking topological states can be characterized by the Chern number, as defined for the two-dimensional partial Brillouin zone. There can be a crossover from a topological insulating to topological semimetal phase in a strong Zeeman fields, and the topological insulator in a Zeeman field is then an example of a symmetry-breaking state. A non-Abelian SU(2) gauge potential has been used to model the spin-orbital coupling in a topological semimetal which was produced by a magnetic field having a π-flux per plaquette and which acted on fermions in a three-dimensional cubic lattice. The Abelian π-flux term gave rise to a spectrum which was characterized by Weyl points. The non-Abelian term was designed to be gauge-equivalent to a two-dimensional Rashba and to a Dresselhaus spin-orbital coupling. Because of the anisotropic characteristics of the coupling between spin and momentum and because of the presence of a C4 rotation-symmetry, the Weyl points exhibited a quadratic dispersion along two directions and acted as double monopoles for the Berry curvature when the non-Abelian part was activated[268]. Non-abelian topological superconductors comprise zero-energy Majorana fermions, bound in quantized vortices, as a result of a bulk topology which is characterized by an odd Chern number. In topological semimetals having a single two-band crossing point, all of the gapped superconductors are non-Abelian. On various three-dimensional lattices, topological semimetals of Majorana fermions arise in exactly solvable Kitaev spin models whose ground-states are quantum spin liquids having the gap-less nodal spectra of bulk Majorana fermion excitations[269]. These phases are topologically stable, provided that some discrete symmetries are protected, and are interesting examples of gap-less topological phases existing in interacting spin systems.

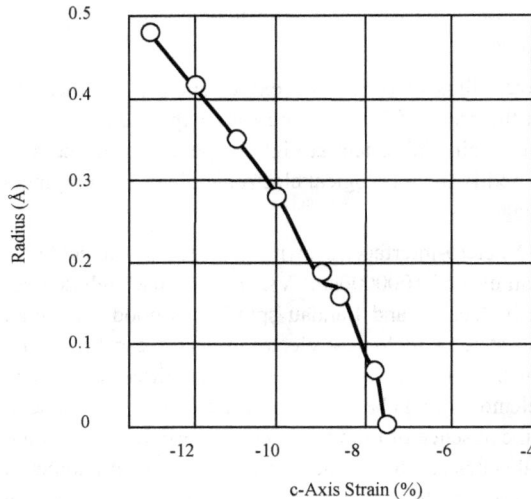

Figure 34. Radius of the nodal ring in YN as a function of uniaxial c-axis strain

Unconventional Hall effects have been observed in a frustrated ferromagnet of pyrochlore $Nd_2Mo_2O_7$, and attributed to finite scalar spin chirality. The resultant Berry phase in real space acts like a magnetic field on conduction electrons: the topological Hall effect. The latter is widely cited as being evidence for skyrmion phases in silicides and germanides, and is indeed a useful means for investigating skyrmion structures in conductive magnetic materials. The topological properties and dynamics of skyrmions have been studied mainly in metals and insulators. Such topological spin structures have been found in oxide films. The ferromagnetic semiconductor, EuO, is noted for containing skyrmionic structures. The topological Hall effect leads to a sharp peak in the magnetization process, in addition to the usual anomalous Hall resistivity which is proportional to the magnetization. But the latter peak occurs only in films which are thinner than 200nm, and is quickly suppressed upon tilting the applied magnetic field away from the surface normal. This behaviour is taken to indicate the formation of two-dimensional skyrmions. This skyrmion is considered to be of Bloch-type which is stabilized in the two-dimensional limit of an Heisenberg ferromagnet. Magnetic phase plotting shows that the two-dimensional skyrmion phase persists down to low temperatures.

Negative magnetoresistance is observed in $PdCoO_2$ when the magnetic field is directed along the interlayer direction, and has been attributed to an axial anomaly between Fermi points in a field-induced quasi one-dimensional dispersion.

Anti-perovskite oxides such as Ca_3SnO, Ca_3PbO, Sr_3SnO, Sr_3PbO, Ba_3SnO and Ba_3PbO are of growing interest due to their Dirac semimetal states. The even simpler rutile-type oxides such as IrO_2, RuO_2 and OsO_2 are expected to be potential nodal topological semimetals. A topological superconducting state with two-dimensional chiral p-wave symmetry is suspected to exist in the layered perovskite, Sr_2RuO_4.

Iridates in particular are useful for research on topological electronic states: the strong spin–orbit interactions lead to low-energy electronic states which are better represented by effective total angular momenta, and the orbitals split into two manifolds. Iridates, epitaxially stabilized on a perovskite substrate, are attracting increasing attention due to the possibly of creating exotic topological phases. Correlated semimetallic electronic structures have been confirmed by means of *in situ* angle-resolved photo-emission spectroscopy. A marked band narrowness arises not only from strong electron correlations and spin–orbit interactions, but also from dimensionality and octahedral rotations of IrO_6. This behaviour is very different to that of less-distorted layered perovskite iridates such as Sr_2IrO_4 and Ba_2IrO_4.

Many pyrochlore iridates, such as $Pr_2Ir_2O_7$ and $Lu_2Ir_2O_7$ and other elements from that group, exhibit an antiferromagnetic spin-ordering in which all of the four spins point inwards or outwards at the vertices of the tetrahedron and are alternately stacked along the [111] direction. Such an antiferromagnetic ordering, with broken time-reversal symmetry, is one required condition for creating a Weyl topological semimetal, with an odd-parity magnetoresistance which depends upon the magnetic field direction.

The above $Pr_2Ir_2O_7$ is metallic down to low temperatures but other pyrochlore iridates undergo a metal-insulator transition. Compounds such as $Nd_2Ir_2O_7$, which are located at the edge of the metal-insulator transition, are therefore expected to be, or near to being, a

Weyl topological semimetal. The same form of spin-ordering has been identified in 5d pyrochlores such as $Cd_2Os_2O_7$, and a similar magnetotransport has been attributed to domain-wall conduction in bulk samples.

Two-dimensional honeycomb lattices have also attracted considerable attention. Thus Na_2IrO_3, which comprises a honeycomb lattice of iridium ions, is a leading example of a two-dimensional topological insulator or quantum spin Hall insulator. Bilayers of perovskite transition metal oxides have been investigated because its lattice structure along the [111]-axis can be viewed as being a honeycomb latticework of the transition metal.

In the case of bulk $Nd_2Ir_2O_7$ polycrystals, an extreme fall in the resistance has been observed upon sweeping the magnetic field, and has been interpreted in terms of conduction occurring at the magnetic domain walls which are randomly formed during the reversal process. Such domain-wall conduction has been confirmed by using a pyrochlore iridate hetero-interface in which one layer was $Eu_2Ir_2O_7$ and the domain pattern was fixed while, in the other layer ($Tb_2Ir_2O_7$), the domain could be reversed by the sweeping field. In this way, single domain walls could be created and annihilated at the interface by sweeping the field, and increased conduction due to domain wall conduction is observed, with a ferroic hysteresis loop. Such pyrochlore iridate hetero-interfaces are thus a new means for investigating the topological surface state.

The topological Hall effect has been observed in an epitaxial heterostructure which combined the itinerant ferromagnet, $SrRuO_3$ with the strong spin–orbit coupling paramagnetic semimetal, $SrIrO_3$. Topological Hall effect resistivity was clearly observed; its magnitude rapidly decreasing with increasing $SrRuO_3$ layer thickness and entirely disappearing when a distance of seven unit cells was exceeded. This suggests that interfacial Dzyaloshinskii–Moriya interaction is needed in order to stabilize nano-scale Néel-type skyrmions in the $SrRuO_3$ layer. High-quality oxide hetero-interfaces clearly provide a basis for the design of topological spin structures.

Ag_2BiO_3

On the basis of first-principles calculations and crystal-symmetry analysis, it is proposed[270] that this centrosymmetric oxide has an hourglass-like nodal net semimetal topology that consists of two hourglass-like nodal chains on mutually orthogonal planes in the extended Brillouin zone, in which weak spin-orbital coupling arises if the 6s-orbital of the bismuth atoms is ignored. The joint point in the nodal net structure is a special double Dirac point which is located at the Brillouin-zone corner. Unlike other hour-glass nodal-net materials in which spin-orbital coupling and double group non-symmorphic symmetries are required and are different to the accidental nodal net, the present structure forms inevitably and is assured by spinless non-symmorphic symmetries. It is thus resistant to the influence of any remaining symmetry-perturbations. The Fermi surface consists of a torus-like electron pocket and a torus-like hole pocket, thus potentially leading to unusual transport behaviour. A simple four-band tight-binding model can be constructed which reproduces the hour-glass nodal-net structure. Assuming a semi-infinite oxide structure, so-called drumhead-like surface states with almost flat dispersions are shown to exist on (001) and (100) surfaces. If a weak spin-orbital coupling is assumed, this hour-glass nodal-net can be slightly broken, leaving a pair of hourglass-like Dirac points at the two-fold screw axis. This type of Dirac semimetal is symmetry-protected and does not require band inversion.

CrO_2

It has been suggested[271] that ferromagnetic topological features such as triple fermions and type-I and type-II Weyl fermions can appear in various forms of this oxide: the rutile form, the orthorhombic form and the pyrite form. First-principles calculations and topological analysis show that breaking of the spin-rotation symmetry produces tunable Weyl fermions that are resistant to spin-orbital coupling. This can be easily achieved by controlling the magnetization direction. Rutile CrO_2, a half-metallic ferromagnet, is an ideal subject for the experimental study of ferromagnetic topological semimetals. It is found that all three forms of the oxide share similar chemical bonds, in that each chromium atom is surrounded by a distorted octahedron of six oxygen atoms; implying that the topological features of the various phases highlight the link between band topology and local chemical bonding.

$CuTeO_3$

First-principles calculations have shown[272] that this monoclinic oxide is a nodal-loop semimetal which has only a single nodal loop around the Fermi level; protected by either PT-symmetry or glide mirror symmetry. Nodal loops can be divided into various classes, depending upon their formation mechanism. Some are accidental, in that their presence requires band-inversion in certain regions of the Brillouin zone and the loop can be adiabatically annihilated without breaking system symmetry. Other nodal loops do not depend upon band inversion. In the absence of spin-orbital coupling, a two-dimensional Z_2 invariant can protect a nodal loop without band inversion. Some non-symmorphic space-group symmetries can guarantee the existence of nodal loops, even in the presence of spin-orbital coupling. The nodal loop in the present material belongs to the first class in that it requires band inversion around the point, and can be annihilated when band-inversion is removed without changing the symmetry of the system. The loop size can be changed by straining and can be entirely eliminated, thus permitting topological transition into a trivial insulator phase. The inclusion of spin-orbital coupling opens up a minute gap in the loop, and the system then becomes a Z_2 topological semimetal with a non-trivial bulk Z_2 invariant but no global band gap.

$Eu_2Ir_2O_7$

Pyrochlore iridates are expected to enjoy a possible topological semimetal status which originates from so-called all-in-all-out spin ordering. The detection of magnetic domain wall conduction offers other interesting possibilities in that two distinct magnetic domains of the all-in-all-out spin structure are known to exhibit linear magnetoresistances but of opposite sign. An investigation[273] of the size of magnetic domains in monocrystalline thin films of the present material, using microscale Hall bars, permitted an estimation to be made of the ratio of the two types of domain in the channel. The linear magnetoresistance of an 80μm x 60μm patterned channel was almost zero following zero-field cooling; thus suggesting the presence of a random distribution of domains which were smaller than the channel size. On the other hand, a wide range of linear-magnetoresistance values was detected in a 2μm x 2μm channel; suggesting that the detectable domain size depended upon the cooling cycle. The average size of a single all-in-all-out magnetic domain was estimated to be 1 to 2μm. Due to the experimental observation of anomalous magnetotransport properties near to the Mott quantum critical point of pyrochlore iridates, a study has been made[274] of the general features of the topological band-structure near to quantum critical points in the presence of a magnetic field. This showed that competition between various energy-scales can generate diverse topological semimetal phases near to quantum critical points. A pivotal factor here is the presence of a quadratic band crossing having four-fold degeneracy in the paramagnetic band-structure. Because of large band-degeneracy and strong spin-orbital coupling, the degenerate states at quadratic band-crossings can exhibit both normal and anisotropic Zeeman effects. Via competition between three different magnetic energy-scales, including the exchange energy between iridium electrons and two Zeeman energies, various topological semimetals can be generated near to a quantum critical point. These three magnetic-energy scales can be affected by modulating the magnetic multipole moment of the cluster of spins within a unit cell. This could couple to the intrinsic magnetic multipole moment of the degenerate states at a quadratic band-crossing.

PbO$_2$

First-principles calculations show[275] that tetragonal, space group P4$_2$/mnm, β-PbO$_2$ (a = 5.079Å, b = 3.446Å) is dynamically unstable at low temperatures due to the existence of a vibrational soft mode, but becomes stable at about 200K, due to the enhanced anharmonic effect. An orthorhombic structure, with a space group of Pnnm (a = 5.054Å, b = 5.098Å, c = 3.446Å), has been identified which is a possible one for the low-temperature ground state of PbO$_2$. Electronic structure calculations suggest that the low-temperature orthorhombic phase is a trivial insulator, whereas the tetragonal high-temperature phase is a topological Dirac nodal line semimetal. This means that a topological phase transition occurs, from a trivial insulator to a topological semimetal, that is linked to the stabilized soft mode and to anharmonicity-driven structural transformation.

SmO

Density functional theory calculations have shown[276] that this correlated mixed-valence oxide is a three-dimensional strongly topological semimetal due to 4f-5d band inversion at the X-point. The conditions under which the inversion is of 4f-6s type make this material quite an unique system. The [001] surface Bloch spectral density reveals two weakly interacting Dirac cones that are quasi-degenerate at the M̄-point and another single Dirac cone at the Γ-point. The topological non-triviality is unaffected upon varying a wide range of lattice parameters, although the electron filling is affected by strain. It has been proposed that it is an ideal test-bed for exploring topological non-trivial correlated flat bands in thin films.

SrIrO$_3$

Gap evolution in t$_{2g}$^{5} perovskite multilayers, (SrTiO$_3$)$_7$/(SrIrO$_3$)$_2$ and (KTaO$_3$)$_7$/(KPtO$_3$)$_2$, has been studied[277] as a function of on-site Coulomb interactions and an imposed uniaxial perpendicular. The molecular beam epitaxial technique is particularly useful for growing high-mobility SrTiO$_3$ films, and the integer quantum Hall effect has been observed in a delta-doped heterostructure. Honeycomb structures, formed by the growth of perovskite transition-metal oxide heterostructures along the (111) direction can produce topological ground-states having a topological index of unity. First-principles calculations of a tight binding could be used to study the multilayers as a function of parity-asymmetry, on-site interaction and uniaxial strain.

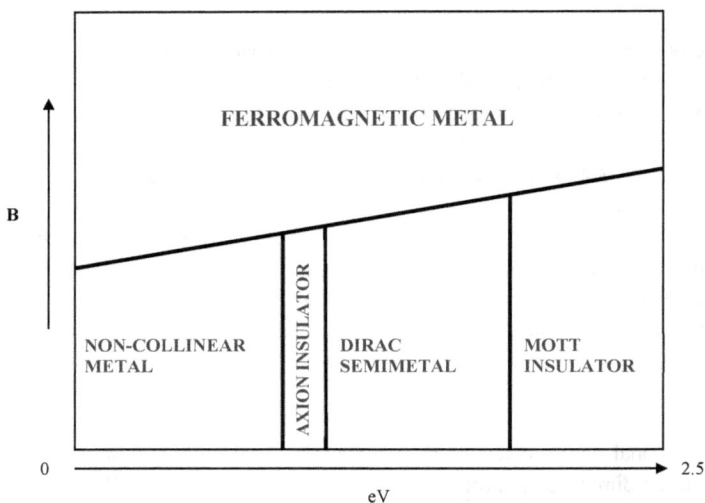

Figure 35. Schematic predicted magnetic phase diagram of pyrochlore iridates
The horizontal scale is a measure of the interaction among 5d iridium electrons, and the
dimensionless vertical axis reflects the magnitude of the external magnetic field

According to density functional theory calculations, only $(SrTiO_3)_7/(SrIrO_3)_2$ is a topological semimetal, while $(KTaO_3)_7/(KPtO_3)_2$ is a topological insulator. In the tight binding model, spin-orbital coupling leads to an effective, $j = \frac{1}{2}$, four-band Hamiltonian. The uniaxial strain and on-site interactions are then trigonal terms in the tight-binding model whose strength controls the size of a topological gap. A comparison of the invariants of the bands indicates that both of these 5d electron systems remain in the intermediate spin-orbit coupling regime. A small gap in the K-point arises from a third-order contribution in the perturbation theory. On the other hand, sub-lattice asymmetry contributes a first-order term, such that the topological phase can be easily destroyed by any external perturbation which produces an inversion-symmetry breaking term in the Hamiltonian. The smallness ($<10meV$) of the gap unfortunately makes the t_{2g}^5 configuration rather unattractive for technological purposes. The system remains interesting because it can be viewed as being an adiabatic deformation of a mathematical model of four-band graphene with spin-orbital coupling and an experimentally measurable gap. More recent work[278] on the effect of flat-surface states on the phonon modes of $SrIrO_3$ side surfaces has shown that mirror odd optical surface phonon modes are different when coupling to surface states having different mirror parities. That is, the first-order self-energy of mirror phonons is such that damping near to the zone center at finite frequencies is zero for even modes but finite for odd modes. This limits the form of electron-phonon interaction for phonons of definite mirror symmetry in some cases. The surface states couple to bulk longitudinal, or surface optical modes with different scalings of the vertex. The damping from surface electrons is different to that due to Landau damping by bulk electrons. It has been shown that the surface dispersions of topological semimetals map onto helicoidal structures when the bulk nodal points are projected onto the branch-points of helicoids whose equal-energy contours are Fermi arcs. When applied to Weyl semimetals, this connection can predict new types of topological semimetals in which the surface states are represented by double- and quad-helicoid surfaces. Each helicoid or multi-helicoid is then a non-compact Riemann surface which represents a multi-valued holomorphic function. The intersection of multiple helicoids, or the branch-cut of the generating function, appears on high-symmetry lines on the surface Brillouin zone; surface states being guaranteed to be doubly-degenerate due to a glide reflection symmetry. It is thus predicted[279] that the heterostructure superlattice, $[(SrIrO_3)_2(CaIrO_3)_2]$, is a topological semimetal having double-helicoid surface states.

$Y_2Ir_2O_7$

The novel phases which arise from the interplay of electron correlations and strong spin-orbital interactions have been studied[280] with regard to topological semimetals: in particular, the pyrochlore iridates. As already explained, this state is a three-dimensional analogue of graphene and provides a condensed-matter manifestation of Weyl fermions that obey a two-component Dirac equation. It also involves surface states, in the form of Fermi arcs, which cannot occur in purely two-dimensional band structures. For intermediate correlation strengths, this is the ground state of pyrochlore iridates co-existing with a non-collinear magnetic order (figure 35). An applied magnetic field can induce a metallic ground state.

Phosphorus and Phosphides

P

Density functional dynamic mean-field theory calculations have shown[281] that gray phosphorus can exhibit a three-dimensional, Kondo semimetal to Fermi liquid phase, cross-over when subjected to an anisotropic compression which affects the in-layer orbital degeneracy. A comparison of the electronic structure and electrical resistivity of black and pressurized gray phosphorus reveals a band-selective Kondo-like electronic reconstruction of the layered p-band allotropes. The 3p spectrum is essentially unaffected in the semiconducting and semimetallic Kondo phases. The Fermi liquid phase however is expected to exhibit a marked electronic reconstruction in the case of compressed gray phosphorus. This observation may aid the understanding of the role of dynamic multi-orbital electronic interactions in the low-energy spectra of topological semimetals. Study of the high magnetic-field transport properties of the Dirac semimetal state of black phosphorus under an applied hydrostatic pressure showed[282] that the band structure undergoes an insulator-semimetal transition. In the high-pressure topological semimetal state, anomalous behaviors of the magnetoresistance and Hall resistivity are predicted to occur, suggesting the existence of an electronic state exhibiting density wave ordering.

Ca3BiP

Pnictide-based anti-perovskites offer the possibility of containing topological phases[283]. The present material exhibits a range of phases when the initial cubic symmetry is gradually broken. An initial small-gap Z_2 topological insulator leads, via spin-orbital coupling, to band re-ordering and finally a topological semimetal phase. Imposition of a compressive uniaxial (001) strain leads back to a small-gap Z_2 topological insulator with its expected gap-less boundary-modes. A tensile (001) strain leaves the system with a pair of Dirac points, along $(0,0,\pm k)$, which pin the Fermi level and produce a novel double-meniscus of connected Fermi arcs on the (100) and (010) surfaces. Breaking of the time-reversal symmetry by applying a Zeeman field produces a new phase having a pair of multi-Weyl nodes. These could be massive or massless, depending upon the direction. They were usually termed semi-Dirac in two-dimensional systems. There was also a pair of Dirac modes along each $\pm k_z$ axis which combined to pin the Fermi level in the immediate vicinity of single-particle excitations. This is just one of a limited number of pnictides of the form, $A_3E^XE^Y$, where A is barium, calcium or strontium and E is antimony, arsenic, bismuth, nitrogen or phosphorus which can be sited on one of two sub-lattices. They are all narrow-gap semiconductors, and a density functional theory-based survey has been made[284] of the entire class of compounds. Upon comparing the relative energetic stabilities of the distribution of pairs of E-ions on the X and Y sites of the structure, it was found that the Y-site always favored the small pnictogen anion. Energy-gap corrections indicated that the cubic structure will furnish just a few topological insulators. Distorted structures, in particular, were expected to lead to materials exhibiting thermoelectric and topological features. A case study of the present compound, which included the effect of strain, demonstrated that a topological semimetal can be transformed into a topological insulator and Dirac semimetal.

CaP_3

Even though it is essentially gap-less in the absence of a magnetic field, theoretical investigations[285] have suggested that an applied magnetic field can create a gap between the conduction and valence bands of a nodal-line semimetal; with the resultant band-gap depending very much upon the strength and orientation of the field. This in turn suggests the existence of a large and variable anisotropy of the magnetoresistance in these topological materials. It has been predicted that some nodal-line semimetals with a single nodal ring, such as the present family, may exhibit a transition between the metallic and insulating states within a magnetic field.

Ca_3P_2

A study was reported of how reflection, time-reversal, SU(2) spin-rotation and inversion symmetries lead to the topological protection of line nodes in three-dimensional semimetals. Crystalline invariants were identified[286] which ensured the stability of line nodes in the bulk, and it was shown that a quantized Berry phase led to the appearance of protected surface states having the form of a drum-head. By deducing a relationship between those crystalline invariants and the Berry phase, a direct connection was found between the stability of line nodes and drum-head surface states. The dispersion minimum of the drum-head state also led to a Van Hove singularity, in the surface density of states, which could serve as the experimental signature of a topological surface state. A typical example of a topological semimetal is the present material, which contains a line of Dirac nodes near to the Fermi energy. The topological properties of this material were explained in terms of a low-energy effective theory and of a tight-binding model which was based upon first-principles density functional theory calculations. The microscopic model showed that the drum-head surface states exhibited quite weak dispersion; implying that correlation effects are greater at the Ca_3P_2 surface. Although some topological semimetals are predicted to exhibit drumhead-like surface states, their direct detection is difficult. It has been proposed[287] that spin-resolved transport in a junction between a normal metal and a spin-orbital coupled nodal-line semimetal could provide a means of detection. In such an interface, drumhead-like surface states introduce a resonant spin-flipped reflection. This can be probed by making vertical spin transport and lateral charge transport measurements between antiparallel magnetic terminals. In the tunnelling limit of the junction, the spin and charge conductances exhibit a resonant peak around zero-energy and provide evidence for the presence of drumhead-like surface states.

MoP

Angle-resolved photo-emission spectroscopy has demonstrated the existence of a triply degenerate point in the electronic structure of this phosphide (figure 36)[288]. This implies the occurrence of quasi-particle excitations, near to a triply degenerate point, which are due to three-component fermions. This goes beyond the usual Dirac-Weyl-Majorana classification, in which Dirac and Weyl fermions are attributed to four-fold and two-fold degenerate points, respectively. Pairs of Weyl points are detected, in the bulk electronic structure, which co-exist with three-component fermions. The material is thus seen as being a test-bed for studies of the interplay between various types of fermion and raises the possibility of its application to topological qubits, spintronics and low-power electronics. Their presence is closely related to the thermal transport properties of the phosphide[289]. Phonon transport was investigated by solving a single-mode relaxation time approximation to the linearized phonon Boltzmann equation. The room-temperature lattice thermal conductivity was calculated to be 18.41 and 34.71W/mK in the in-plane and cross-plane directions, respectively. Isotope scattering had little effect, and phonons made little contribution to the lattice thermal conductivity when the phonon mean free path was greater than 0.15μm at 300K. The average room-temperature lattice thermal conductivity was lower than that of the TaAs Weyl semimetal and this was attributed to lower group-velocities and larger Grüneisen parameters; the average Grüneisen parameter of 1.57 indicating a relatively strong anharmonic phonon scattering.

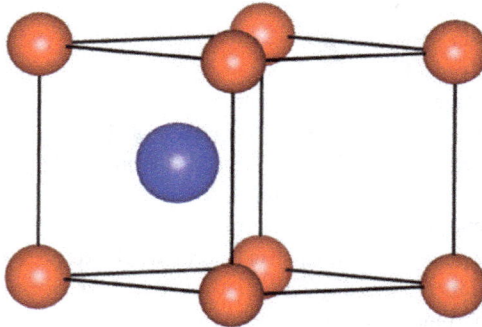

Figure 36. Unit cell of MoP

Again in connection with the coexistence of triply degenerate points of band-crossing and Weyl points near to the Fermi level of this material, high-pressure electrical transport measurements revealed the onset of a pressure-induced superconductivity having a critical transition temperature of about 2.5K at around 30GPa (figure 37)[290]. Synchrotron X-ray diffraction studies did not detect any structural phase transition at up to about 60GPa. It was therefore deduced that a topologically non-trivial band which was protected by crystalline symmetries, and superconductivity, could co-exist at pressures above 30GPa; in agreement with density functional theory predictions. This makes the pressurized material a promising topological superconductor. Even the high normal conductivity of this phosphide is useful in speeding up charge transport between a catalyst and electrolyte, and electrolytic processes such as hydrogen evolution require highly efficient catalysts having great surface stability. The phosphide was encapsulated in a (Mo,P)-codoped carbon layer, and this composite exhibited remarkable hydrogen evolution; with the extremely low over-potential of 49mV at a current density of 10mA/cm^2 and a Tafel slope of 54mV/dec in an alkaline medium[291]. Electron transport analysis showed that the material exhibits a high conductivity and mobility due to the above triple-point fermions and complex Fermi surface. The presence of P-C and Mo-C bonds at the interface between the carbon layer and phosphide particles affected the band structure of the composite and aided rapid electron transfer, accumulation and delocalization; again ensuring excellent hydrogen evolution.

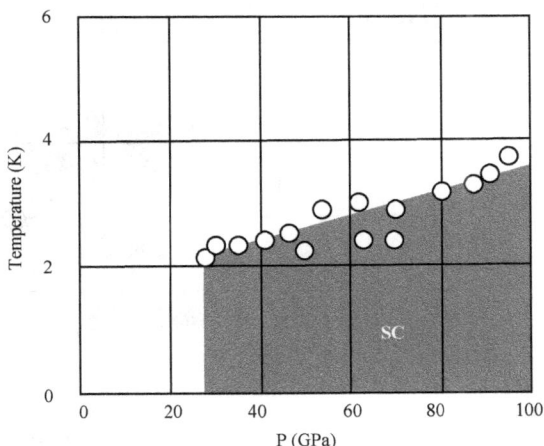

Figure 37. Pressure–temperature phase diagram of MoP
SC: superconducting

Equally interesting are the properties of MoP_2: a type-II Weyl semimetal having robust Weyl points[292]. Single crystals exhibit an extremely low residual low-temperature ($3n\Omega cm$), together with a very high and very anisotropic magnetoresistance of above 2 x 10^8% at 63T and 2.5K (figure 38). There is a marked suppression of charge carrier back-scattering.

The hydrogen-evolution-reaction electrolytic process requires highly efficient catalysts having great surface stability and a high conductivity. Such a conductivity is exhibited by MoP, making it a model catalyst for this purpose[293]. When encapsulated in a (Mo,P)-doped carbon layer, it has the very low over-potential of 49mV at a current density of $10mA/cm^2$, and a Tafel slope of 54mV/dec in alkaline media. Electron-transport analysis indicates that the high conductivity and mobility are due to the existence of triple-point fermions and a complex Fermi surface. The existence of P-C and Mo-C bonds at the carbon-layer/MoP- particle interface modulates the band structure of the composite and facilitates electron transfer, accumulation and subsequent delocalization.

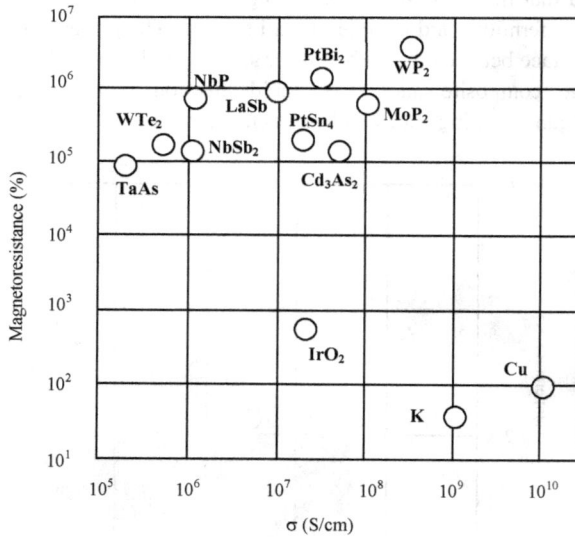

Figure 38. Magnetoresistance and conductivity of MoP₂, at 2K and in 9T, as compared with other materials. Note that high-conductivity metals exhibit a smaller magnetoresistance and lower-conductivity semimetals exhibit a higher magnetoresistance

NbP

This semimetal exhibits a large non-saturating magnetoresistance, and has a band structure which is predicted to combine the characteristics of a Weyl semimetal with those of a normal semimetal[294]. A magnetoresistance of 850000% is observed at 1.85K, and of 250% at room temperature, in a field of up to 9T, with no sign of saturation. There is also a carrier mobility of $5 \times 10^6 \mathrm{cm}^2/\mathrm{Vs}$, which is accompanied by strong Shubnikov-de Haas oscillations. The bulk polycrystalline material has a markedly higher Nernst thermopower, than its conventional thermopower, in a magnetic field: a maximum Nernst power factor of about $35 \times 10^{-4} \mathrm{W/mK}^2$ being found[295] in a field of 9T at 136K. It retains a relatively large value over a wide range of temperatures. Density functional theory calculations, and an analysis of angle- and temperature-dependent quantum oscillations, indicated[296] that there are coexisting p- and n-type Weyl pockets in the planes near to high-symmetry points. One of the pockets formed a large dumbbell-shaped Fermi surface which encloses two neighboring Weyl nodes of opposite chirality. Magnetotransport is dominated by these highly anisotropic pockets, where the Weyl fermions are protected from defect back-scattering by real spin conservation associated with the chiral nodes. Upon doping with just 1% of chromium, the mobility is degraded by more than two orders of magnitude, due to a loss of helicity protection. Helicity-protected Weyl fermion transport is also reflected by a chiral-anomaly induced negative magnetoresistance, controlled by another Weyl state. In the quantum regime below 10K, the intervalley scattering-time due to impurities becomes large and constant and leads to a sudden conductivity increase in low magnetic fields.

Table 7. Formation energies of various defects in TaP

Defect	Energy (eV)
Interstitial hydrogen	-3.13
Tantalum vacancy	6.07
Phosphorus vacancy	5.99
Hydrogen at tantalum site	0.35
Hydrogen at phosphorus site	0.50

TaP

This material has been predicted to be a topological semimetal of Weyl type, in which spin-polarized band-crossings, the Weyl nodes, are connected by topological surface arcs. Low-energy excitations near to the crossing-points here behave like massless Weyl fermions and are expected to exhibit unusual properties, such as a chiral anomaly. In order for the transport properties to be dominated by Weyl fermions, the Weyl nodes have to be located nearly to the chemical potential and to be enclosed by pairs of individual Fermi surfaces having non-zero Fermi Chern numbers. A combination of angle-resolved photo-emission spectroscopy and first-principles calculations showed[297] that it is a Weyl semimetal having only a single type of Weyl fermion. This differed from the case of TaAs, where two types of Weyl fermion contribute to the low-energy physical properties. Defects which had been introduced by hydrogen ion bombardment (table 7) were studied[298] by using density functional theory calculations, and the various defects were found to exert differing effects upon the topological properties. It was noted that Weyl-point positions were not affected by most of the defects. Hydrogen atoms, present as interstitials, could modify the Fermi level. The presence of substitutional hydrogen atoms at phosphorus sites did not affect the topological properties. On the other hand, phosphorus and tantalum vacancies at a concentration of 1/64, and substitutional hydrogen atoms at tantalum sites, destroyed some of the Weyl points.

Many causes of the extreme non-saturating magnetoresistance have been proposed, but the predominant mechanism operating in a specific material can be unclear. A study[299] of the magnetic susceptibility, tangent of the Hall angle and magnetoresistance of high-mobility non-magnetic semimetals, NbP, TaP, $NbSb_2$ and $TaSb_2$, has shown that the distinctly different temperature dependences of those properties can be used as criteria for identifying the magnetoresistance contributions arising from various sources. In particular, NbP and TaP are uncompensated semimetals with linear dispersion and here the non-saturating magnetoresistance arises from guiding center motion. On the other hand, $NbSb_2$ and $TaSb_2$ are compensated semimetals in which the magnetoresistance arises due to the almost perfect charge compensation of two quadratic bands.

WP₂

Angle-resolved photo-emission spectroscopy and density functional theory calculations[300] have been combined in order to investigate a possible relationship between an extremely large magnetoresistance and the presence of Weyl points, the aim being to demarcate the surface and bulk contributions to the spectroscopic intensity. It was shown that, although the hole-like and electron-like Fermi surface sheets which originate from surface states have differing areas, the bulk band structure of this material is electron-hole compensated; in agreement with density functional theory. The spectroscopic band structure is compatible with the existence of at least four temperature-independent Weyl points; thus confirming the topological nature of the material, and its stability with respect to lattice distortion.

Silicides

Al_3FeSi_2

Nodal-line semimetals have a degeneracy along nodal lines where the band-gap is closed. The nodal lines may appear accidentally and it is then impossible to decide whether the nodal line appeared due to the details of crystal symmetry and electron filling. In the case of spinless systems, it has been shown[301] that for certain space groups with 4N+2 filling, or 8N+4 filling when the spin degree of freedom is included, the presence of nodal lines is required, regardless of the system details. Spinless systems are here crystals for which spin-orbital coupling is negligible and the spin degree of freedom can be ignored because of SU(2) spin degeneracy. In this case the shape of the band-structure around these nodal lines is like an hourglass; consequently leading to a so-called spinless hour-glass nodal-line semimetal. Construction of a model Hamiltonian showed that it is always in the spinless hour-glass nodal-line semimetal phase even when the model parameters are changed without changing the system symmetries. A list of all of the centrosymmetric space groups for which spinless systems always have hour-glass nodal lines was established, and showed where the nodal lines are located. It was suggested that the present material, with its Pbcn space-group symmetry, was one of the nodal-line semimetals which was governed by the above mechanism.

Ba_2Si

First-principles calculations show[302] that Ba_2X, where X is silicon or germanium, exhibits a topological semimetal phase having one nodal ring in the $k_x = 0$ plane which is protected by glide mirror symmetry when spin-orbital coupling is ignored. The corresponding drumhead-like surface flat band appears on the (100) surface in Green's function calculations. A topological semimetal-to-insulator transition is also identified. The nodal-line semimetal is expected to change into a topological insulator when spin-orbital coupling is incorporated. Topologically-protected metallic surface states appear around the Γ-point and could be turned into topologically trivial insulator states by applying more than 3% of hydrostatic strain. These materials thus exhibit a quantum phase transition, between a topological semimetal and an insulator, which can be effected by elastic straining.

CoSi

This topological semimetal, and its solid solutions with iron or nickel, have a B20 cubic non-centrosymmetric structure and exhibit interesting thermoelectric properties. First-principles calculations of the scattering of charge carriers by phonons and point defects have shown[303] that the dependence of the scattering rate upon energy is related to that on the total density of states; suggesting that not only intraband but also interband scattering is important, especially for bands with a low density of states. Calculations of the Seebeck coefficient and electrical resistivity of CoSi and of dilute iron or nickel solid solutions show that a marked energy-dependence of the relaxation time is important in describing an experimentally observed rapid increase in the resistivity and a qualitative change in the temperature dependence upon substituting cobalt for iron. Point-defect scattering and phonon scattering of charge carriers is greater in alloys with iron than in those with nickel, due to the higher scattering potential and larger density of states at the Fermi level. An observed decrease in the resistivity, with temperature, of iron-doped material was due to an increase in the contribution of carriers having a higher energy and greater mobility.

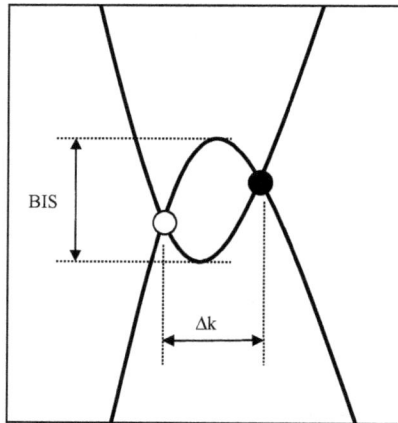

Figure 39. Schematic pair of band-inversion Weyl nodes (gapless bulk crossing points) in a crystal without mirror symmetry. The white and black circles correspond to Weyl nodes of positive and negative chirality, respectively

Co₂TiSi

Magnetic topological semimetal states have been identified[304] in ferromagnetic half-metal compounds of the form, Co_2TiX, having Curie temperatures greater than 350K where X is silicon, germanium or tin. First-principles band-structure calculations show that the compounds have three topological nodal lines in the absence of spin-orbital coupling. The inclusion of spin-orbital coupling then gives rise to Weyl nodes whose momentum-space locations can be controlled by changing the magnetization direction.

RhSi

It has been suggested that this compound is an ideal test-bed for the experimental study of unusual chiral fermions. It is a filling-enforced semimetal with, located near to its Fermi surface, a chiral doubly six-fold degenerate spin-1 Weyl node at the R-point and a fourfold-degenerate chiral fermion at the Γ-point. Each of the unusual fermions has a Chern number of ±4 at the Fermi level[305]. This material also exhibited the largest possible momentum separation of compensating chiral fermions, the widest topologically non-trivial energy-window and the longest possible Fermi arcs on the surface. In a structurally chiral system where only a single two-band Weyl fermion is partially unoccupied, such as a Kramers Weyl metal, the difference in the rate of current density which results from exciting electrons with left-hand and right-hand circularly polarized light is quantized. Here, the four-fold fermion at the Γ-point is located just above the Fermi energy while the chiral double spin-1 Weyl node at the R-point is located below and is fully occupied. Angular-momentum selection rules for circularly polarized light appear to limit strongly the permissible transitions of this four-fold fermion, so that only transitions between certain bands contribute to the photocurrent.

SrSi₂

A first-principles study[306], based upon hybrid exchange-correlation functionals, has been made of the topological electronic structure of this, and analogous, non-centrosymmetric materials. The topological strength of Weyl semimetal states is determined by the band inversion strength (BIS) and the separation (Δk) of Weyl fermions of opposite chiral charge. (figure 39). The topological phases could be mapped (figure 40) as a function of the lattice constant. The material comprises a pair of double-Weyl nodes having a chiral charge of ±2, the separation between the double-Weyl nodes being small. The material lies close to a topological semimetal-to-insulator transition point. An adjustable double-Weyl fermion state was identified in $Sr_{1-x}Ca_xSi_2$ and $Sr_{1-x}Ba_xSi_2$ alloys. Calcium-doping was shown to yield a double-Weyl semimetal having a large Fermi arc-length, whereas barium-doping led to a transition from topological semimetal to gapped insulator.

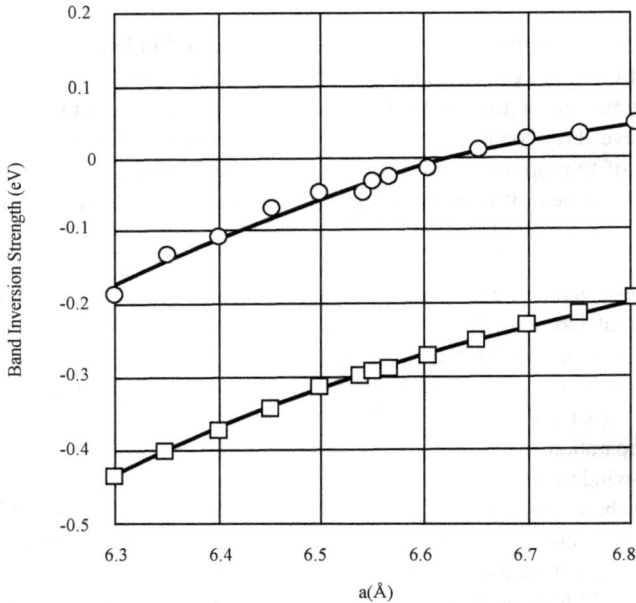

*Figure 40. Band inversion strength of SrSi₂ as a function of the lattice constant
Circles: Heyd-Scuseria-Ernzerof exchange-correlation functional,
squares: generalized-gradient-approximation*

Miscellaneous

GdSI

Materials of the type, LnSI, where Ln is gadolinium or lutetium, in honeycomb-lattice form can exhibit ideal Weyl semimetal, three-dimensional strong topological insulator and nodal-line semimetal configurations. It has been shown[307] that GdSI, an ideal Weyl semimetal, has two pairs of Weyl nodes which reside at the Fermi level. Meanwhile, density functional theory calculations predict that LuSI and YSI should be strong three-dimensional topological insulators having unusual right-handed helical Dirac cones. Two very long Fermi-arcs exist on the (010) surface of GdSI.

References

[1] Bachmann, M.D., Nair, N., Flicker, F., Ilan, R., Meng, T., Ghimire, N.J., Bauer, E.D., Ronning, F., Analytis, J.G., Moll, P.J.W., Science Advances, 3[5] 2017, e1602983. https://doi.org/10.1126/sciadv.1602983

[2] Politano, A., Chiarello, G., Li, Z., Fabio, V., Wang, L., Guo, L., Chen, X., Boukhvalov, D.W., Advanced Functional Materials, 28[23] 2018, 1800511. https://doi.org/10.1002/adfm.201800511

[3] Rajamathi, C.R., Gupta, U., Kumar, N., Yang, H., Sun, Y., Süß, V., Shekhar, C., Schmidt, M., Blumtritt, H., Werner, P., Yan, B., Parkin, S., Felser, C., Rao, C.N.R., Advanced Materials, 29[19] 2017, 1606202. https://doi.org/10.1002/adma.201606202

[4] Hills, R.D.Y., Kusmartseva, A., Kusmartsev, F.V., Physical Review B, 95[21] 2017, 214103. https://doi.org/10.1103/PhysRevB.95.214103

[5] Guo, Q., Yang, B., Xia, L., Gao, W., Liu, H., Chen, J., Xiang, Y., Zhang, S., Physical Review Letters, 119[21] 2017, 213901. https://doi.org/10.1103/PhysRevLett.119.213901

[6] Sun, K., Liu, W.V., Hemmerich, A., Das Sarma, S., Nature Physics, 8[1] 2012, 67-70.

[7] Hou, J.M., Physical Review Letters, 111[13] 2013, 130403. https://doi.org/10.1103/PhysRevLett.111.130403

[8] Bednik, G., Zyuzin, A.A., Burkov, A.A., New Journal of Physics, 18[8] 2016, 085002. https://doi.org/10.1088/1367-2630/18/8/085002

[9] Wang, R., Wang, B., Shen, R., Sheng, L., Xing, D.Y., EPL, 105[1] 2014, 17004. https://doi.org/10.1209/0295-5075/105/17004

[10] Po, H.C., Bahri, Y., Vishwanath, A., Physical Review B, 93[20] 2016, 205158. https://doi.org/10.1103/PhysRevB.93.205158

[11] Ge, H., Ni, X., Tian, Y., Gupta, S.K., Lu, M.H., Lin, X., Huang, W.D., Chan, C.T., Chen, Y.F., Physical Review Applied, 10[1] 2018, 014017. https://doi.org/10.1103/PhysRevApplied.10.014017

[12] Volovik, G.E., Zubkov, M.A., New Journal of Physics, 19[1] 2017, 015009. https://doi.org/10.1088/1367-2630/aa573d

[13] Ho, C.H., Chang, C.P., Lin, M.F., Physical Review B, 93[7] 2016, 075437. https://doi.org/10.1103/PhysRevB.93.075437

[14] Hyart, T., Ojajärvi, R., Heikkilä, T.T., Journal of Low Temperature Physics, 191[1-2] 2018, 35-48. https://doi.org/10.1007/s10909-017-1846-3

[15] Tan, X., Zhao, Y., Liu, Q., Xue, G., Yu, H., Wang, Z.D., Yu, Y., npj Quantum Materials, 2[1] 2017, 60.

[16] Grushin, A.G., Venderbos, J.W.F., Vishwanath, A., Ilan, R., Physical Review X, 6[4] 2016, 041046. https://doi.org/10.1103/PhysRevX.6.041046

[17] Hayami, S., Motome, Y., Physical Review B, 91[7] 2015, 075104. https://doi.org/10.1103/PhysRevB.91.075104

[18] Ahn, J., Kim, D., Kim, Y., Yang, B.J., Physical Review Letters, 121[10] 2018, 106403. https://doi.org/10.1103/PhysRevLett.121.106403

[19] Mathai, V., Thiang, G.C., Communications in Mathematical Physics, 355[2] 2017, 561-602. https://doi.org/10.1007/s00220-017-2965-z

[20] Weyl, H., Zeitschrift für Physik, 56[5–6] 1929, 330-352. https://doi.org/10.1007/BF01339504

[21] Wallace, P.R., Physical Review, 71[9] 1947, 622-634. https://doi.org/10.1103/PhysRev.71.622

[22] Wallace, P.R., Physical Review, 72, 1947, 258. https://doi.org/10.1103/PhysRev.72.258

[23] Ramamurthy, S.T., Hughes, T.L., Physical Review B, 92[8] 2015, 085105. https://doi.org/10.1103/PhysRevB.92.085105

[24] Kim, Y., Kang, K., Schleife, A., Gilbert, M.J., Physical Review B, 97[13] 2018, 134415. https://doi.org/10.1103/PhysRevB.97.134415

[25] Ma, Q., Xu, S.Y., Chan, C.K., Zhang, C.L., Chang, G., Lin, Y., Xie, W., Palacios, T., Lin, H., Jia, S., Lee, P.A., Jarillo-Herrero, P., Gedik, N., Nature Physics, 13[9] 2017, 842-847.

[26] Guan, S., Yu, Z.M., Liu, Y., Liu, G.B., Dong, L., Lu, Y., Yao, Y., Yang, S.A., npj Quantum Materials, 2[1] 2017, 26.

[27] Murakami, S., Hirayama, M., Okugawa, R., Miyake, T., Science Advances, 3[5] 2017, e1602680. https://doi.org/10.1126/sciadv.1602680

[28] Watanabe, H., Po, H.C., Zaletel, M.P., Vishwanath, A., Physical Review Letters, 117[9] 2016, 096404. https://doi.org/10.1103/PhysRevLett.117.096404

[29] Ono, S., Watanabe, H., Physical Review B, 98[11] 2018, 115150.

https://doi.org/10.1103/PhysRevB.98.115150

[30] Zhang, J., Chan, Y.H., Chiu, C.K., Vergniory, M.G., Schoop, L.M., Schnyder, A.P., Physical Review Materials, 2[7] 2018, 074201. https://doi.org/10.1103/PhysRevMaterials.2.074201

[31] Watanabe, H., Po, H.C., Vishwanath, A., Science Advances, 4[8] 2018, eaat8685.

[32] Gautier, R., Zhang, X., Hu, L., Yu, L., Lin, Y., Sunde, T.O.L., Chon, D., Poeppelmeier, K.R., Zunger, A., Nature Chemistry, 7[4] 2015, 308-316. https://doi.org/10.1038/nchem.2207

[33] Ezawa, M., Physical Review B, 96[4] 2017, 041202. https://doi.org/10.1103/PhysRevB.96.041202

[34] Gong, C., Xie, Y., Chen, Y., Kim, H.S., Vanderbilt, D., Physical Review Letters, 120[10] 2018, 106403. https://doi.org/10.1103/PhysRevLett.120.106403

[35] Zhou, Y., Xiong, F., Wan, X., An, J., Physical Review B, 97[15] 2018, 155140. https://doi.org/10.1103/PhysRevB.97.155140

[36] Song, Z., Zhang, T., Fang, C., Physical Review X, 8[3] 2018, 031069. https://doi.org/10.1103/PhysRevX.8.031069

[37] Chu, R.L., Shan, W.Y., Lu, J., Shen, S.Q., Physical Review B, 83[7] 2011, 075110. https://doi.org/10.1103/PhysRevB.83.075110

[38] Zhong, C., Chen, Y., Xie, Y., Yang, S.A., Cohen, M.L., Zhang, S.B., Nanoscale, 8[13] 2016, 7232-7239. https://doi.org/10.1039/C6NR00882H

[39] Slager, R.J., Juričić, V., Lahtinen, V., Zaanen, J., Physical Review B, 93[24] 2016, 245406. https://doi.org/10.1103/PhysRevB.93.245406

[40] Wang, J.T., Nie, S., Weng, H., Kawazoe, Y., Chen, C., Physical Review Letters, 120[2] 2018, 026402. https://doi.org/10.1103/PhysRevLett.120.026402

[41] Wang, G., Xu, H., Lai, Y.C., Physical Review B, 95[23] 2017, 235159. https://doi.org/10.1103/PhysRevB.95.235159

[42] Yue, S.Y., Qin, G., Zhang, X., Sheng, X., Su, G., Hu, M., Physical Review B, 95[8] 2017, 085207. https://doi.org/10.1103/PhysRevB.95.085207

[43] Lin, M., Hughes, T.L., Physical Review B, 98[24] 2018, 241103. https://doi.org/10.1103/PhysRevB.98.241103

[44] Ezawa, M., Physical Review B, 97[15] 2018, 155305. https://doi.org/10.1103/PhysRevB.97.155305

[45] Lu, H.Z., Shen, S.Q., Frontiers of Physics, 12[3] 2017, 127201.
https://doi.org/10.1007/s11467-016-0609-y

[46] Dai, X., Lu, H.Z., Shen, S.Q., Yao, H., Physical Review B, 93[16] 2016, 161110.
https://doi.org/10.1103/PhysRevB.93.161110

[47] Lu, H.Z., Zhang, S.B., Shen, S.Q., Physical Review B, 92[4] 2015, 045203.
https://doi.org/10.1103/PhysRevB.92.045203

[48] Wang, J., Liu, Y., Jin, K.H., Sui, X., Zhang, L., Duan, W., Liu, F., Huang, B.,
Physical Review B, 98[20] 2018, 201112.
https://doi.org/10.1103/PhysRevB.98.201112

[49] Yang, Y., Bai, C., Xu, X., Jiang, Y., Nanotechnology, 29[7] 2018, 074002.
https://doi.org/10.1088/1361-6528/aaa0bd

[50] Zhao, X.M., Kong, X., Guo, C.X., Wu, Y.J., Kou, S.P., EPL, 120[4] 2017, 47004.
https://doi.org/10.1209/0295-5075/120/47004

[51] Vazifeh, M.M., EPL, 102[6] 2013, 67011. https://doi.org/10.1209/0295-
5075/102/67011

[52] Burkov, A.A., Physical Review B, 97[16] 2018, 165104.
https://doi.org/10.1103/PhysRevB.97.165104

[53] Nandy, S., Taraphder, A., Tewari, S., Scientific Reports, 8[1] 2018, 14983.
https://doi.org/10.1038/s41598-018-33258-5

[54] Zhang, S.B., Lu, H.Z., Shen, S.Q., New Journal of Physics, 18[5] 2016, 053039.
https://doi.org/10.1088/1367-2630/18/5/053039

[55] Kim, P., Ryoo, J.H., Park, C.H., Physical Review Letters, 119[26] 2017, 266401.
https://doi.org/10.1103/PhysRevLett.119.266401

[56] Jin, Y.J., Wang, R., Chen, Z.J., Zhao, J.Z., Zhao, Y.J., Xu, H., Physical Review B,
96[20] 2017, 201102. https://doi.org/10.1103/PhysRevB.96.201102

[57] Liu, W.E., Hankiewicz, E.M., Culcer, D., Physical Review B, 96[4] 2017, 045307.
https://doi.org/10.1103/PhysRevB.96.045307

[58] Yesilyurt, C., Siu, Z.B., Tan, S.G., Liang, G., Jalil, M.B.A., Journal of Applied
Physics, 121[24] 2017, 244303. https://doi.org/10.1063/1.4989993

[59] Grignani, G., Marini, A., Pe-a-Benitez, F., Speziali, S., Journal of High Energy
Physics, 2017[3] 2017, 125.

[60] Chiu, P.M., Huang, C.Y., Li, W.J., Lee, T.K., Journal of Physics - Condensed

Matter, 31[3] 2019, 035501. https://doi.org/10.1088/1361-648X/aaf040

[61] Ganeshan, S., Das Sarma, S., Physical Review B, 91[12] 2015, 125438. https://doi.org/10.1103/PhysRevB.91.125438

[62] Mathai, V., Thiang, G.C., Journal of Physics A, 50[11] 2017, 11LT01.

[63] Fang, C., Chen, Y., Kee, H.Y., Fu, L., Physical Review B, 92[8] 2015, 081201. https://doi.org/10.1103/PhysRevB.92.081201

[64] Chen, W., Lu, H.Z., Hou, J.M., Physical Review B, 96[4] 2017, 041102. https://doi.org/10.1103/PhysRevB.96.041102

[65] Bi, R., Yan, Z., Lu, L., Wang, Z., Physical Review B, 96[20] 2017, 201305. https://doi.org/10.1103/PhysRevB.96.201305

[66] Yan, Z., Bi, R., Shen, H., Lu, L., Zhang, S.C., Wang, Z., Physical Review B, 96[4] 2017, 041103. https://doi.org/10.1103/PhysRevB.96.041103

[67] Martín-Ruiz, A., Cortijo, A., Physical Review B, 98[15] 2018, 155125. https://doi.org/10.1103/PhysRevB.98.155125

[68] He, J., Kong, X., Wang, W., Kou, S.P., New Journal of Physics, 20[5] 2018, 053019. https://doi.org/10.1088/1367-2630/aabdf8

[69] Lim, L.K., Moessner, R., Physical Review Letters, 118[1] 2017, 016401. https://doi.org/10.1103/PhysRevLett.118.016401

[70] Wu, W., Liu, Y., Li, S., Zhong, C., Yu, Z.M., Sheng, X.L., Zhao, Y.X., Yang, S.A., Physical Review B, 97[11] 2018, 115125. https://doi.org/10.1103/PhysRevB.97.115125

[71] Liu, Y., Sun, Y.W., Journal of High Energy Physics, 12, 2018, 72. https://doi.org/10.1007/JHEP12(2018)072

[72] Yang, B.J., Bojesen, T.A., Morimoto, T., Furusaki, A., Physical Review B, 95[7] 2017, 075135. https://doi.org/10.1103/PhysRevB.95.075135

[73] Ramamurthy, S.T., Hughes, T.L., Physical Review B, 95[7] 2017, 075138. https://doi.org/10.1103/PhysRevB.95.075138

[74] Wawrzik, D., Lindner, D., Hermanns, M., Trebst, S., Physical Review B, 98[11] 2018, 115114. https://doi.org/10.1103/PhysRevB.98.115114

[75] Gresch, D., Autès, G., Yazyev, O.V., Troyer, M., Vanderbilt, D., Bernevig, B.A., Soluyanov, A.A., Physical Review B, 95[7] 2017, 075146. https://doi.org/10.1103/PhysRevB.95.075146

[76] Lee, W.R., Park, K., Physical Review B, 92[19] 2015, 195144.
https://doi.org/10.1103/PhysRevB.92.195144

[77] Parameswaran, S.A., Grover, T., Abanin, D.A., Pesin, D.A., Vishwanath, A.,
Physical Review X, 4[3] 2014, 031035.
https://doi.org/10.1103/PhysRevX.4.031035

[78] Dai, X., Du, Z.Z., Lu, H.Z., Physical Review Letters, 119[16] 2017, 166601.
https://doi.org/10.1103/PhysRevLett.119.166601

[79] McCormick, T.M., Watzman, S.J., Heremans, J.P., Trivedi, N., Physical Review B,
97[19] 2018, 195152. https://doi.org/10.1103/PhysRevB.97.195152

[80] Hu, Y., Liu, H., Jiang, H., Xie, X.C., Physical Review B, 96[13] 2017, 134201.
https://doi.org/10.1103/PhysRevB.96.134201

[81] Fisher, D., Materials Research Foundations, 14, 2017, 1-372

[82] Ares, P., Palacios, J.J., Abellán, G., Gómez-Herrero, J., Zamora, F., Advanced
Materials, 30[2] 2018, 1703771. https://doi.org/10.1002/adma.201703771

[83] Soumyanarayanan, A., Hoffman, J.E., Journal of Electron Spectroscopy and Related
Phenomena, 201, 2015, 66-73. https://doi.org/10.1016/j.elspec.2014.10.008

[84] Zhang, P., Liu, Z., Duan, W., Liu, F., Wu, J., Physical Review B, 85[20] 2012,
201410. https://doi.org/10.1103/PhysRevB.85.201410

[85] Wang, X., Bian, G., Xu, C., Wang, P., Hu, H., Zhou, W., Brown, S.A., Chiang, T.C.,
Nanotechnology, 28[39] 2017, 395706. https://doi.org/10.1088/1361-6528/aa825f

[86] Pakdel, S., Pourfath, M., Palacios, J.J., Beilstein Journal of Nanotechnology, 9[1]
2018, 1015-1023. https://doi.org/10.3762/bjnano.9.94

[87] Vincent, T., Vlaic, S., Pons, S., Zhang, T., Aubin, H., Stolyarov, V.S., Ksenz, A.S.,
Ionov, A.M., Chekmazov, S.V., Bozhko, S.I., Roditchev, D., Physical Review B,
98[15] 2018, 155440. https://doi.org/10.1103/PhysRevB.98.155440

[88] Ye, L., Suzuki, T., Wicker, C.R., Checkelsky, J.G., Physical Review B, 97[8] 2018,
081108. https://doi.org/10.1103/PhysRevB.97.081108

[89] Pardo, V., Smith, J.C., Pickett, W.E., Physical Review B, 85[21] 2012, 214531.
https://doi.org/10.1103/PhysRevB.85.214531

[90] Winkler, G.W., Wu, Q., Troyer, M., Krogstrup, P., Soluyanov, A.A., Physical
Review Letters, 117[7] 2016, 076403.
https://doi.org/10.1103/PhysRevLett.117.076403

[91] Tafti, F.F., Gibson, Q.D., Kushwaha, S.K., Haldolaarachchige, N., Cava, R.J., Nature Physics, 12[3] 2016, 272-277.

[92] Pavlosiuk, O., Kleinert, M., Swatek, P., Kaczorowski, D., Wiśniewski, P., Scientific Reports, 7[1] 2017, 12822. https://doi.org/10.1038/s41598-017-12792-8

[93] Kawasaki, J.K., Sharan, A., Johansson, L.I.M., Hjort, M., Timm, R., Thiagarajan, B., Schultz, B.D., Mikkelsen, A., Janotti, A., Palmstrøm, C.J., Science Advances, 4[6] 2018, eaar5832.

[94] Guo, L., Liu, Y.K., Gao, G.Y., Huang, Y.Y., Gao, H., Chen, L., Zhao, W., Ren, W., Li, S.Y., Li, X.G., Dong, S., Zheng, R.K., Journal of Applied Physics, 123[15] 2018, 155103. https://doi.org/10.1063/1.5021637

[95] Wakeham, N., Bauer, E.D., Neupane, M., Ronning, F., Physical Review B, 93[20] 2016, 205152. https://doi.org/10.1103/PhysRevB.93.205152

[96] Wang, Y., Yu, J.H., Wang, Y.Q., Xi, C.Y., Ling, L.S., Zhang, S.L., Wang, J.R., Xiong, Y.M., Han, T., Han, H., Yang, J., Gong, J., Luo, L., Tong, W., Zhang, L., Qu, Z., Han, Y.Y., Zhu, W.K., Pi, L., Wan, X.G., Zhang, C., Zhang, Y., Physical Review B, 97[11] 2018, 115133. https://doi.org/10.1103/PhysRevB.97.115133

[97] Liu, J.Y., Hu, J., Zhang, Q., Graf, D., Cao, H.B., Radmanesh, S.M.A., Adams, D.J., Zhu, Y.L., Cheng, G.F., Liu, X., Phelan, W.A., Wei, J., Jaime, M., Balakirev, F., Tennant, D.A., DItusa, J.F., Chiorescu, I., Spinu, L., Mao, Z.Q., Nature Materials, 16[9] 2017, 905-910. https://doi.org/10.1038/nmat4953

[98] Li, Y., Li, L., Wang, J., Wang, T., Xu, X., Xi, C., Cao, C., Dai, J., Physical Review B, 94[12] 2016, 121115. https://doi.org/10.1103/PhysRevB.94.121115

[99] Liu, X.Y., Wang, J.L., You, W., Wang, T.T., Yang, H.Y., Jiao, W.H., Mao, H.Y., Zhang, L., Cheng, J., Li, Y.K., Chinese Physics Letters, 34[12] 2017, 127501. https://doi.org/10.1088/0256-307X/34/12/127501

[100] Pariari, A., Singha, R., Roy, S., Satpati, B., Mandal, P., Scientific Reports, 8[1] 2018, 10527. https://doi.org/10.1038/s41598-018-28922-9

[101] Zhou, Y., Gu, C., Chen, X., Zhou, Y., An, C., Yang, Z., Journal of Solid State Chemistry, 265, 2018, 359-363. https://doi.org/10.1016/j.jssc.2018.06.027

[102] Pavlosiuk, O., Swatek, P.A., Wiśniewski, P., Scientific Reports, 6, 2016, 38691. https://doi.org/10.1038/srep38691

[103] Li, T., Sushkov, O.P., Physical Review B, 96[8] 2017, 085301. https://doi.org/10.1103/PhysRevB.96.085301

[104] Emmanouilidou, E., Shen, B., Deng, X., Chang, T.R., Shi, A., Kotliar, G., Xu, S.Y., Ni, N., Physical Review B, 95[24] 2017, 245113. https://doi.org/10.1103/PhysRevB.95.245113

[105] Takane, D., Nakayama, K., Souma, S., Wada, T., Okamoto, Y., Takenaka, K., Yamakawa, Y., Yamakage, A., Mitsuhashi, T., Horiba, K., Kumigashira, H., Takahashi, T., Sato, T., npj Quantum Materials, 3[1] 2018, 1.

[106] Lin-Chung, P.J., Physical Review, 188[3] 1969, 1272-1280. https://doi.org/10.1103/PhysRev.188.1272

[107] Caron, L.G., Jay-Gerin, J.P., Aubin, M.J., Physical Review, 15[8] 1977, 3879-3887. https://doi.org/10.1103/PhysRevB.15.3879

[108] Wang, Z., Weng, H., Wu, Q., Dai, X., Fang, Z., Physical Review B, 88[12] 2013, 125427. https://doi.org/10.1103/PhysRevB.88.125427

[109] Yi, H., Wang, Z., Chen, C., Shi, Y., Feng, Y., Liang, A., Xie, Z., He, S., He, J., Peng, Y., Liu, X., Liu, Y., Zhao, L., Liu, G., Dong, X., Zhang, J., Nakatake, M., Arita, M., Shimada, K., Namatame, H., Taniguchi, M., Xu, Z., Chen, C., Dai, X., Fang, Z., Zhou, X.J., Scientific Reports, 4, 2014, 6106. https://doi.org/10.1038/srep06106

[110] Neupane, M., Xu, S.Y., Sankar, R., Alidoust, N., Bian, G., Liu, C., Belopolski, I., Chang, T.R., Jeng, H.T., Lin, H., Bansil, A., Chou, F., Hasan, M.Z., Nature Communications, 5, 2014, 3786. https://doi.org/10.1038/ncomms4786

[111] Wu, D.S., Wang, X., Zhang, X., Yang, C.L., Zheng, P., Li, P.G., Shi, Y.G., Science China: Physics, Mechanics and Astronomy, 58[1] 2014, 1-6. https://doi.org/10.1007/s11433-014-5608-9

[112] Li, C.Z., Li, J.G., Wang, L.X., Zhang, L., Zhang, J.M., Yu, D., Liao, Z.M., ACS Nano, 10[6] 2016, 6020-6028. https://doi.org/10.1021/acsnano.6b01568

[113] Li, H., He, H., Lu, H.Z., Zhang, H., Liu, H., Ma, R., Fan, Z., Shen, S.Q., Wang, J., Nature Communications, 7, 2016, 10301. https://doi.org/10.1038/ncomms10301

[114] Schumann, T., Goyal, M., Kealhofer, D.A., Stemmer, S., Physical Review B, 95[24] 2017, 241113. https://doi.org/10.1103/PhysRevB.95.241113

[115] Zhang, C., Zhang, E., Wang, W., Liu, Y., Chen, Z.G., Lu, S., Liang, S., Cao, J., Yuan, X., Tang, L., Li, Q., Zhou, C., Gu, T., Wu, Y., Zou, J., Xiu, F., Nature Communications, 8, 2017, 13741. https://doi.org/10.1038/ncomms13741

[116] Zhou, T., Zhang, C., Zhang, H., Xiu, F., Yang, Z., Inorganic Chemistry Frontiers,

3[12] 2016, 1637-1643. https://doi.org/10.1039/C6QI00383D

[117] Zhang, C., Narayan, A., Lu, S., Zhang, J., Zhang, H., Ni, Z., Yuan, X., Liu, Y., Park, J.H., Zhang, E., Wang, W., Liu, S., Cheng, L., Pi, L., Sheng, Z., Sanvito, S., Xiu, F., Nature Communications, 8[1] 2017, 1272. https://doi.org/10.1038/s41467-017-01438-y

[118] Wang, C.M., Lu, H.Z., Shen, S.Q., Physical Review Letters, 117[7] 2016, 077201. https://doi.org/10.1103/PhysRevLett.117.077201

[119] Sharafeev, A., Gnezdilov, V., Sankar, R., Chou, F.C., Lemmens, P., Physical Review B, 95[23] 2017, 235148. https://doi.org/10.1103/PhysRevB.95.235148

[120] Conte, A.M., Pulci, O., Bechstedt, F., Scientific Reports, 7, 2017, 45500. https://doi.org/10.1038/srep45500

[121] Schumann, T., Galletti, L., Kealhofer, D.A., Kim, H., Goyal, M., Stemmer, S., Physical Review Letters, 120[1] 2018, 016801. https://doi.org/10.1103/PhysRevLett.120.016801

[122] Goryunov, Y.V., Nateprov, A.N., Physics of the Solid State, 60[1] 2018, 68-74. https://doi.org/10.1134/S1063783418010109

[123] Li, C.Z., Li, C., Wang, L.X., Wang, S., Liao, Z.M., Brinkman, A., Yu, D.P., Physical Review B, 97[11] 2018, 115446. https://doi.org/10.1103/PhysRevB.97.115446

[124] Nakazawa, Y., Uchida, M., Nishihaya, S., Kriener, M., Kozuka, Y., Taguchi, Y., Kawasaki, M., Scientific Reports, 8[1] 2018, 2244. https://doi.org/10.1038/s41598-018-20758-7

[125] Wang, S., Lin, B.C., Zheng, W.Z., Yu, D., Liao, Z.M., Physical Review Letters, 120[25] 2018, 257701. https://doi.org/10.1103/PhysRevLett.120.257701

[126] Uykur, E., Sankar, R., Schmitz, D., Kuntscher, C.A., Physical Review B, 97[19] 2018, 195134. https://doi.org/10.1103/PhysRevB.97.195134

[127] Kargarian, M., Lu, Y.M., Randeria, M., Physical Review B, 97[16] 2018, 165129. https://doi.org/10.1103/PhysRevB.97.165129

[128] Zhang, X., Sun, S., Lei, H., Physical Review B, 96[23] 2017, 235105. https://doi.org/10.1103/PhysRevB.96.235105

[129] Emmanouilidou, E., Liu, J., Graf, D., Cao, H., Ni, N., Journal of Magnetism and Magnetic Materials, 469, 2019, 570-573. https://doi.org/10.1016/j.jmmm.2018.08.084

[130] Wang, G., Wei, J., Computational Materials Science, 124, 2016, 311-315. https://doi.org/10.1016/j.commatsci.2016.08.005

[131] Barik, R.K., Shinde, R., Singh, A.K., Journal of Physics - Condensed Matter, 30[37] 2018, 375702. https://doi.org/10.1088/1361-648X/aad8e1

[132] Lu, J.L., Luo, W., Li, X.Y., Yang, S.Q., Cao, J.X., Gong, X.G., Xiang, H.J., Chinese Physics Letters, 34[5] 2017, 057302. https://doi.org/10.1088/0256-307X/34/5/057302

[133] Zhang, F., Zhou, J., Xiao, D., Yao, Y., Physical Review Letters, 119[26] 2017, 266804. https://doi.org/10.1103/PhysRevLett.119.266804

[134] Khalid, S., Sabino, F.P., Janotti, A., Physical Review B, 98[22] 2018, 220102. https://doi.org/10.1103/PhysRevB.98.220102

[135] Xu, S.Y., Alidoust, N., Belopolski, I., Yuan, Z., Bian, G., Chang, T.R., Zheng, H., Strocov, V.N., Sanchez, D.S., Chang, G., Zhang, C., Mou, D., Wu, Y., Huang, L., Lee, C.C., Huang, S.M., Wang, B., Bansil, A., Jeng, H.T., Neupert, T., Kaminski, A., Lin, H., Jia, S., Hasan, M.Z., Nature Physics, 11[9] 2015, 748-754.

[136] Weng, H., Fang, C., Fang, Z., Andrei Bernevig, B., Dai, X., Physical Review X, 5[1] 2015, 011029. https://doi.org/10.1103/PhysRevX.5.011029

[137] Lee, C.C., Xu, S.Y., Huang, S.M., Sanchez, D.S., Belopolski, I., Chang, G., Bian, G., Alidoust, N., Zheng, H., Neupane, M., Wang, B., Bansil, A., Hasan, M.Z., Lin, H., Physical Review B, 92[23] 2015, 235104. https://doi.org/10.1103/PhysRevB.92.235104

[138] Sun, Y., Wu, S.C., Yan, B., Physical Review B, 92[11] 2015, 115428. https://doi.org/10.1103/PhysRevB.92.115428

[139] Li, Y., Wang, Z., Li, P., Yang, X., Shen, Z., Sheng, F., Li, X., Lu, Y., Zheng, Y., Xu, Z.A., Frontiers of Physics, 12[3] 2017, 127205. https://doi.org/10.1007/s11467-016-0636-8

[140] Khouri, T., Zeitler, U., Reichl, C., Wegscheider, W., Hussey, N.E., Wiedmann, S., Maan, J.C., Physical Review Letters, 117[25] 2016, 256601. https://doi.org/10.1103/PhysRevLett.117.256601

[141] Yuan, X., Yan, Z., Song, C., Zhang, M., Li, Z., Zhang, C., Liu, Y., Wang, W., Zhao, M., Lin, Z., Xie, T., Ludwig, J., Jiang, Y., Zhang, X., Shang, C., Ye, Z., Wang, J., Chen, F., Xia, Z., Smirnov, D., Chen, X., Wang, Z., Yan, H., Xiu, F., Nature Communications, 9[1] 2018, 1854.

[142] Grassano, D., Pulci, O., Mosca Conte, A., Bechstedt, F., Scientific Reports, 8[1] 2018, 3534. https://doi.org/10.1038/s41598-018-21465-z

[143] Moll, P.J.W., Potter, A.C., Nair, N.L., Ramshaw, B.J., Modic, K.A., Riggs, S., Zeng, B., Ghimire, N.J., Bauer, E.D., Kealhofer, R., Ronning, F., Analytis, J.G., Nature Communications, 7, 2016, 12492. https://doi.org/10.1038/ncomms12492

[144] Luo, Y., Ghimire, N.J., Bauer, E.D., Thompson, J.D., Ronning, F., Journal of Physics Condensed Matter, 28[5] 2016, 055502. https://doi.org/10.1088/0953-8984/28/5/055502

[145] Zhang, J., Liu, F.L., Dong, J.K., Xu, Y., Li, N.N., Yang, W.G., Li, S.Y., Chinese Physics Letters, 32[9] 2015, 097102. https://doi.org/10.1088/0256-307X/32/9/097102

[146] Gupta, S.N., Singh, A., Pal, K., Muthu, D.V.S., Shekhar, C., Elghazali, M.A., Naumov, P.G., Medvedev, S.A., Felser, C., Waghmare, U.V., Sood, A.K., Journal of Physics - Condensed Matter, 30[18] 2018, 185401. https://doi.org/10.1088/1361-648X/aab5e3

[147] Guo, Z.P., Lu, P.C., Chen, T., Wu, J.F., Sun, J., Xing, D.Y., Science China - Physics, Mechanics and Astronomy, 61[3] 2018, 038211. https://doi.org/10.1007/s11433-017-9126-6

[148] Wang, C.M., Sun, H.P., Lu, H.Z., Xie, X.C., Physical Review Letters, 119[13] 2017, 136806. https://doi.org/10.1103/PhysRevLett.119.136806

[149] Chang, G., Xu, S.Y., Zheng, H., Lee, C.C., Huang, S.M., Belopolski, I., Sanchez, D.S., Bian, G., Alidoust, N., Chang, T.R., Hsu, C.H., Jeng, H.T., Bansil, A., Lin, H., Hasan, M.Z., Physical Review Letters, 116[6] 2016, 066601. https://doi.org/10.1103/PhysRevLett.116.066601

[150] Lv, B.Q., Xu, N., Weng, H.M., Ma, J.Z., Richard, P., Huang, X.C., Zhao, L.X., Chen, G.F., Matt, C.E., Bisti, F., Strocov, V.N., Mesot, J., Fang, Z., Dai, X., Qian, T., Shi, M., Ding, H., Nature Physics, 11[9] 2015, 724-727.

[151] Gyenis, A., Inoue, H., Jeon, S., Zhou, B.B., Feldman, B.E., Wang, Z., Li, J., Jiang, S., Gibson, Q.D., Kushwaha, S.K., Krizan, J.W., Ni, N., Cava, R.J., Bernevig, B.A., Yazdani, A., New Journal of Physics, 18[10] 2016, 105003. https://doi.org/10.1088/1367-2630/18/10/105003

[152] Dadsetani, M., Ebrahimian, A., Journal of Electronic Materials, 45[11] 2016, 5867-5876. https://doi.org/10.1007/s11664-016-4766-0

[153] Yan, B., Felser, C., Annual Review of Condensed Matter Physics, 8, 2017, 337-354. https://doi.org/10.1146/annurev-conmatphys-031016-025458

[154] Guo, C., Tian, H.F., Yang, H.X., Sun, K., Wei, L.L., Chen, G.F., Li, J.Q., Crystal Growth and Design, 17[4] 2017, 1747-1751. https://doi.org/10.1021/acs.cgd.6b01743

[155] Luo, Y., McDonald, R.D., Rosa, P.F.S., Scott, B., Wakeham, N., Ghimire, N.J., Bauer, E.D., Thompson, J.D., Ronning, F., Scientific Reports, 6, 2016, 27294. https://doi.org/10.1038/srep27294

[156] Wang, Y.Y., Yu, Q.H., Guo, P.J., Liu, K., Xia, T.L., Physical Review B, 94[4] 2016, 041103. https://doi.org/10.1103/PhysRevB.94.041103

[157] Peramaiyan, G., Sankar, R., Muthuselvam, I.P., Lee, W.L., Scientific Reports, 8[1] 2018, 6414. https://doi.org/10.1038/s41598-018-24823-z

[158] Weber, S.F., Chen, R., Yan, Q., Neaton, J.B., Physical Review B, 96[23] 2017, 235145. https://doi.org/10.1103/PhysRevB.96.235145

[159] Li, Y., Xu, C., Shen, M., Wang, J., Yang, X., Yang, X., Zhu, Z., Cao, C., Xu, Z.A., Physical Review B, 98[11] 2018, 115145. https://doi.org/10.1103/PhysRevB.98.115145

[160] Shekhar, C., Kumar, N., Grinenko, V., Singh, S., Sarkar, R., Luetkens, H., Wu, S.C., Zhang, Y., Komarek, A.C., Kampert, E., Skourski, Y., Wosnitza, J., Schnelle, W., McCollam, A., Zeitler, U., Kübler, J., Yan, B., Klauss, H.H., Parkin, S.S.P., Felser, C., Proceedings of the National Academy of Sciences of the United States of America, 115[37] 2018, 9140-9144. https://doi.org/10.1073/pnas.1810842115

[161] Nikitin, A.M., Pan, Y., Mao, X., Jehee, R., Araizi, G.K., Huang, Y.K., Paulsen, C., Wu, S.C., Yan, B.H., De Visser, A., Journal of Physics - Condensed Matter, 27[27] 2015, 275701. https://doi.org/10.1088/0953-8984/27/27/275701

[162] Pavlosiuk, O., Kaczorowski, D., Fabreges, X., Gukasov, A., Wisniewski, P., Scientific Reports, 6, 2016, 18797. https://doi.org/10.1038/srep18797

[163] Nakajima, Y., Hu, R., Kirshenbaum, K., Hughes, A., Syers, P., Wang, X., Wang, K., Wang, R., Saha, S.R., Pratt, D., Lynn, J.W., Paglione, J., Science Advances, 1[5] 2015, e1500242. https://doi.org/10.1126/sciadv.1500242

[164] Pavlosiuk, O., Fabreges, X., Gukasov, A., Meven, M., Kaczorowski, D., Wiśniewski, P., Physica B, 536, 2018, 56-59.

https://doi.org/10.1016/j.physb.2017.10.062

[165] Ekahana, S.A., Wu, S.C., Jiang, J., Okawa, K., Prabhakaran, D., Hwang, C.C., Mo, S.K., Sasagawa, T., Felser, C., Yan, B., Liu, Z., Chen, Y., New Journal of Physics, 19[6] 2017, 065007. https://doi.org/10.1088/1367-2630/aa75a1

[166] Zhao, Y.X., Schnyder, A.P., Physical Review B, 94[19] 2016, 195109. https://doi.org/10.1103/PhysRevB.94.195109

[167] Le, C., Qin, S., Wu, X., Dai, X., Fu, P., Fang, C., Hu, J., Physical Review B, 96[11] 2017, 115121. https://doi.org/10.1103/PhysRevB.96.115121

[168] Zhang, X., Sun, S., Lei, H., Physical Review B, 95[3] 2017, 035209. https://doi.org/10.1103/PhysRevB.95.035209

[169] Tafti, F.F., Torikachvili, M.S., Stillwell, R.L., Baer, B., Stavrou, E., Weir, S.T., Vohra, Y.K., Yang, H.Y., McDonnell, E.F., Kushwaha, S.K., Gibson, Q.D., Cava, R.J., Jeffries, J.R., Physical Review B, 95[1] 2017, 014507. https://doi.org/10.1103/PhysRevB.95.014507

[170] Tafti, F.F., Gibson, Q., Kushwaha, S., Krizan, J.W., Haldolaarachchige, N., Cava, R.J., Proceedings of the National Academy of Sciences of the United States of America, 113[25] 2016, E3475-E3481. https://doi.org/10.1073/pnas.1607319113

[171] Wu, Y., Kong, T., Wang, L.L., Johnson, D.D., Mou, D., Huang, L., Schrunk, B., Bud'Ko, S.L., Canfield, P.C., Kaminski, A., Physical Review B, 94[8] 2016, 081108. https://doi.org/10.1103/PhysRevB.94.081108

[172] Kumar, N., Shekhar, C., Wu, S.C., Leermakers, I., Young, O., Zeitler, U., Yan, B., Felser, C., Physical Review B, 93[24] 2016, 241106. https://doi.org/10.1103/PhysRevB.93.241106

[173] Kumar, N., Shekhar, C., Klotz, J., Wosnitza, J., Felser, C., Physical Review B, 96[16] 2017, 161103. https://doi.org/10.1103/PhysRevB.96.161103

[174] Cheng, X., Li, R., Li, D., Li, Y., Chen, X.Q., Physical Review B, 92[15] 2015, 155109. https://doi.org/10.1103/PhysRevB.92.155109

[175] Nie, T., Meng, L., Li, Y., Luan, Y., Yu, J., Journal of Physics - Condensed Matter, 30[12] 2018, 125502. https://doi.org/10.1088/1361-648X/aaad22

[176] Han, S., Moon, E.G., Physical Review B, 97[24] 2018, 241101. https://doi.org/10.1103/PhysRevB.97.241101

[177] Dadsetani, M., Ebrahimian, A., Journal of Physics and Chemistry of Solids, 100, 2017, 161-169. https://doi.org/10.1016/j.jpcs.2016.10.002

[178] Iwaya, K., Kohsaka, Y., Okawa, K., Machida, T., Bahramy, M.S., Hanaguri, T., Sasagawa, T., Nature Communications, 8[1] 2017, 1209. https://doi.org/10.1038/s41467-017-01303-y

[179] Gao, W., Hao, N., Zheng, F.W., Ning, W., Wu, M., Zhu, X., Zheng, G., Zhang, J., Lu, J., Zhang, H., Xi, C., Yang, J., Du, H., Zhang, P., Zhang, Y., Tian, M., Physical Review Letters, 118[25] 2017, 256601. https://doi.org/10.1103/PhysRevLett.118.256601

[180] Butch, N.P., Syers, P., Kirshenbaum, K., Hope, A.P., Paglione, J., Physical Review B, 84[22] 2011, 220504. https://doi.org/10.1103/PhysRevB.84.220504

[181] Meinert, M., Physical Review Letters, 116[13] 2016, 137001. https://doi.org/10.1103/PhysRevLett.116.137001

[182] Kim, H., Wang, K., Nakajima, Y., Hu, R., Ziemak, S., Syers, P., Wang, L., Hodovanets, H., Denlinger, J.D., Brydon, P.M.R., Agterberg, D.F., Tanatar, M.A., Prozorov, R., Paglione, J., Science Advances, 4[4] 2018, eaao4513.

[183] Nourbakhsh, Z., Faizi-Mohazzab, B., Journal of Magnetism and Magnetic Materials, 396, 2015, 106-112. https://doi.org/10.1016/j.jmmm.2015.08.007

[184] Ding, G., Gao, G.Y., Yu, L., Ni, Y., Yao, K., Journal of Applied Physics, 119[2] 2016, 025105. https://doi.org/10.1063/1.4939887

[185] Guo, C., Wu, F., Smidman, M., Yuan, H., AIP Advances, 8[10] 2018, 101336. https://doi.org/10.1063/1.5043049

[186] Li, D., Chen, Y., He, J., Tang, Q., Zhong, C., Ding, G., Chinese Physics B, 27[3] 2018, 036303. https://doi.org/10.1088/1674-1056/27/3/036303

[187] Zabolotskiy, A.D., Lozovik, Y.E., Physical Review B, 94[16] 2016, 165403. https://doi.org/10.1103/PhysRevB.94.165403

[188] Cheng, T., Lang, H., Li, Z., Liu, Z., Liu, Z., Physical Chemistry Chemical Physics, 19[35] 2017, 23942-23950. https://doi.org/10.1039/C7CP03736H

[189] Nakhaee, M., Ketabi, S.A., Peeters, F.M., Physical Review B, 98[11] 2018, 115413. https://doi.org/10.1103/PhysRevB.98.115413

[190] Fan, X., Ma, D., Fu, B., Liu, C.C., Yao, Y., Physical Review B, 98[19] 2018, 195437. https://doi.org/10.1103/PhysRevB.98.195437

[191] Gupta, S., Kutana, A., Yakobson, B.I., Journal of Physical Chemistry Letters, 9[11] 2018, 2757-2762. https://doi.org/10.1021/acs.jpclett.8b00640

Materials Research Forum LLC
doi: http://dx.doi.org/10.21741/9781644900154

[192] Xu, R., Zou, X., Liu, B., Cheng, H.M., Materials Today, 21[4] 2018, 391-418. https://doi.org/10.1016/j.mattod.2018.03.003

[193] Liu, J., Wu, J., Chen, C., Han, L., Zhu, Z., Wu, J., International Journal of Modern Physics B, 32[4] 2018, 1850033. https://doi.org/10.1142/S0217979218500339

[194] Wang, Q., Guo, P.J., Sun, S., Li, C., Liu, K., Lu, Z.Y., Lei, H., Physical Review B, 97[20] 2018, 205105. https://doi.org/10.1103/PhysRevB.97.205105

[195] Kauppila, V.J., Hyart, T., Heikkilä, T.T., Physical Review B, 93[2] 2016, 024505. https://doi.org/10.1103/PhysRevB.93.024505

[196] Hyart, T., Heikkilä, T.T., Physical Review B, 93[23] 2016, 235147. https://doi.org/10.1103/PhysRevB.93.235147

[197] Wang, J.T., Chen, C., Kawazoe, Y., Physical Review B, 97[24] 2018, 245147. https://doi.org/10.1103/PhysRevB.97.245147

[198] Wang, J.T., Weng, H., Nie, S., Fang, Z., Kawazoe, Y., Chen, C., Physical Review Letters, 116[19] 2016, 195501. https://doi.org/10.1103/PhysRevLett.116.195501

[199] Cheng, Y., Feng, X., Cao, X., Wen, B., Wang, Q., Kawazoe, Y., Jena, P., Small, 13[12] 2017, 1602894. https://doi.org/10.1002/smll.201602894

[200] Liu, J., Wang, S., Sun, Q., Proceedings of the National Academy of Sciences of the United States of America, 114[4] 2017, 651-656. https://doi.org/10.1073/pnas.1618051114

[201] Zhou, P., Ma, Z.S., Sun, L.Z., Journal of Materials Chemistry C, 6[5] 2018, 1206-1214. https://doi.org/10.1039/C7TC05095J

[202] Sun, J.P., Zhang, D., Chang, K., Chinese Physics Letters, 34[2] 2017, 027102. https://doi.org/10.1088/0256-307X/34/2/027102

[203] Khazaei, M., Ranjbar, A., Arai, M., Yunoki, S., Physical Review B, 94[12] 2016, 125152. https://doi.org/10.1103/PhysRevB.94.125152

[204] He, J.B., Chen, D., Zhu, W.L., Zhang, S., Zhao, L.X., Ren, Z.A., Chen, G.F., Physical Review B, 95[19] 2017, 195165. https://doi.org/10.1103/PhysRevB.95.195165

[205] Guo, S.D., Chen, P., Journal of Chemical Physics, 148[14] 2018, 144706. https://doi.org/10.1063/1.5026644

[206] Ma, J.Z., He, J.B., Xu, Y.F., Lv, B.Q., Chen, D., Zhu, W.L., Zhang, S., Kong, L.Y., Gao, X., Rong, L.Y., Huang, Y.B., Richard, P., Xi, C.Y., Choi, E.S., Shao, Y.,

Wang, Y.L., Gao, H.J., Dai, X., Fang, C., Weng, H.M., Chen, G.F., Qian, T., Ding, H., Nature Physics, 14[4] 2018, 349-354.

[207] Li, J., Xie, Q., Ullah, S., Li, R., Ma, H., Li, D., Li, Y., Chen, X.Q., Physical Review B, 97[5] 2018, 054305. https://doi.org/10.1103/PhysRevB.97.054305

[208] Hirayama, M., Matsuishi, S., Hosono, H., Murakami, S., Physical Review X, 8[3] 2018, 031067. https://doi.org/10.1103/PhysRevX.8.031067

[209] Niu, C., Buhl, P.M., Bihlmayer, G., Wortmann, D., Dai, Y., Blügel, S., Mokrousov, Y., 2017 Physical Review B, 95[23], 235138. https://doi.org/10.1103/PhysRevB.95.235138

[210] Valla, T., Ji, H., Schoop, L.M., Weber, A.P., Pan, Z.H., Sadowski, J.T., Vescovo, E., Fedorov, A.V., Caruso, A.N., Gibson, Q.D., Müchler, L., Felser, C., Cava, R.J., Physical Review B, 86[24] 2012, 241101. https://doi.org/10.1103/PhysRevB.86.241101

[211] Sheng, X.L., Yu, Z.M., Yu, R., Weng, H., Yang, S.A., Journal of Physical Chemistry Letters, 8[15] 2017, 3506-3511. https://doi.org/10.1021/acs.jpclett.7b01390

[212] Kumar, N., Manna, K., Qi, Y., Wu, S.C., Wang, L., Yan, B., Felser, C., Shekhar, C., Physical Review B, 95[12] 2017, 121109. https://doi.org/10.1103/PhysRevB.95.121109

[213] Rauch, T., Achilles, S., Henk, J., Mertig, I., Physical Review Letters, 114[23] 2015, 236805. https://doi.org/10.1103/PhysRevLett.114.236805

[214] Du, Y., Bo, X., Wang, D., Kan, E.J., Duan, C.G., Savrasov, S.Y., Wan, X., Physical Review B, 96[23] 2017, 235152. https://doi.org/10.1103/PhysRevB.96.235152

[215] Sun, J.P., Chinese Physics Letters, 34[7] 2017, 077101. https://doi.org/10.1088/0256-307X/34/7/077101

[216] Roy, S., Pariari, A., Singha, R., Satpati, B., Mandal, P., Applied Physics Letters, 112[16] 2018, 162402. https://doi.org/10.1063/1.5024479

[217] Matusiak, M., Cooper, J.R., Kaczorowski, D., Nature Communications, 8, 2017, 15219. https://doi.org/10.1038/ncomms15219

[218] Hu, J., Tang, Z., Liu, J., Zhu, Y., Wei, J., Mao, Z., Physical Review B, 96[4] 2017, 045127. https://doi.org/10.1103/PhysRevB.96.045127

[219] Pezzini, S., Van Delft, M.R., Schoop, L.M., Lotsch, B.V., Carrington, A.,

Katsnelson, M.I., Hussey, N.E., Wiedmann, S., Nature Physics, 14[2] 2018, 178-183.

[220] Zhang, C.L., Schindler, F., Liu, H., Chang, T.R., Xu, S.Y., Chang, G., Hua, W., Jiang, H., Yuan, Z., Sun, J., Jeng, H.T., Lu, H.Z., Lin, H., Hasan, M.Z., Xie, X.C., Neupert, T., Jia, S., Physical Review B, 96[16] 2017, 165148. https://doi.org/10.1103/PhysRevB.96.165148

[221] Weber, A.P., Gibson, Q.D., Ji, H., Caruso, A.N., Fedorov, A.V., Cava, R.J., Valla, T., Physical Review Letters, 114[25] 2015, 256401. https://doi.org/10.1103/PhysRevLett.114.256401

[222] Xu, G., Weng, H., Wang, Z., Dai, X., Fang, Z., Physical Review Letters, 107[18] 2011, 186806. https://doi.org/10.1103/PhysRevLett.107.186806

[223] Fang, C., Gilbert, M.J., Dai, X., Bernevig, B.A., Physical Review Letters, 108[26] 2012, 266802. https://doi.org/10.1103/PhysRevLett.108.266802

[224] Lin, C., Yi, C., Shi, Y., Zhang, L., Zhang, G., Müller, J., Li, Y., Physical Review B, 94[22] 2016, 224404. https://doi.org/10.1103/PhysRevB.94.224404

[225] Lin, C.J., Shi, Y.G., Li, Y.Q., Chinese Physics Letters, 33[7] 2016, 077501. https://doi.org/10.1088/0256-307X/33/7/077501

[226] Bian, G., Chang, T.R., Sankar, R., Xu, S.Y., Zheng, H., Neupert, T., Chiu, C.K., Huang, S.M., Chang, G., Belopolski, I., Sanchez, D.S., Neupane, M., Alidoust, N., Liu, C., Wang, B., Lee, C.C., Jeng, H.T., Zhang, C., Yuan, Z., Jia, S., Bansil, A., Chou, F., Lin, H., Hasan, M.Z., Nature Communications, 7, 2016, 10556. https://doi.org/10.1038/ncomms10556

[227] Zhu, Z., Chang, T.R., Huang, C.Y., Pan, H., Nie, X.A., Wang, X.Z., Jin, Z.T., Xu, S.Y., Huang, S.M., Guan, D.D., Wang, S., Li, Y.Y., Liu, C., Qian, D., Ku, W., Song, F., Lin, H., Zheng, H., Jia, J.F., Nature Communications, 9[1] 2018, 4153. https://doi.org/10.1038/s41467-018-06661-9

[228] Li, C., Wang, C.M., Wan, B., Wan, X., Lu, H.Z., Xie, X.C., Physical Review Letters, 120[14] 2018, 146602. https://doi.org/10.1103/PhysRevLett.120.146602

[229] Hu, J., Zhu, Y.L., Graf, D., Tang, Z.J., Liu, J.Y., Mao, Z.Q., Physical Review B, 95[20] 2017, 205134. https://doi.org/10.1103/PhysRevB.95.205134

[230] Schoop, L.M., Topp, A., Lippmann, J., Orlandi, F., Müchler, L., Vergniory, M.G., Sun, Y., Rost, A.W., Duppel, V., Krivenkov, M., Sheoran, S., Manuel, P., Varykhalov, A., Yan, B., Kremer, R.K., Ast, C.R., Lotsch, B.V., Science

Advances, 4[2] 2018, eaar2317.

[231] Pesin, D.A., Nature Materials, 17[9] 2018, 750-751.
https://doi.org/10.1038/s41563-018-0161-y

[232] Kim, K., Seo, J., Lee, E., Ko, K.T., Kim, B.S., Jang, B.G., Ok, J.M., Lee, J., Jo, Y.J., Kang, W., Shim, J.H., Kim, C., Yeom, H.W., Il Min, B., Yang, B.J., Kim, J.S., Nature Materials, 17[9] 2018, 794-799. https://doi.org/10.1038/s41563-018-0132-3

[233] Crepaldi, A., Roth, S., Gatti, G., Arrell, C.A., Ojeda, J., Van Mourik, F., Bugnon, P., Magrez, A., Berger, H., Chergui, M., Grioni, M., Chimia, 71[5] 2017, 273-277. https://doi.org/10.2533/chimia.2017.273

[234] Hosen, M.M., Dhakal, G., Dimitri, K., Maldonado, P., Aperis, A., Kabir, F., Sims, C., Riseborough, P., Oppeneer, P.M., Kaczorowski, D., Durakiewicz, T., Neupane, M., Scientific Reports, 8[1] 2018, 13283. https://doi.org/10.1038/s41598-018-31296-7

[235] Wang, Z., Gresch, D., Soluyanov, A.A., Xie, W., Kushwaha, S., Dai, X., Troyer, M., Cava, R.J., Bernevig, B.A., Physical Review Letters, 117[5] 2016, 056805. https://doi.org/10.1103/PhysRevLett.117.056805

[236] Cho, S., Kang, S.H., Yu, H.S., Kim, H.W., Ko, W., Hwang, S.W., Han, W.H., Choe, D.H., Jung, Y.H., Chang, K.J., Lee, Y.H., Yang, H., Kim, S.W., 2D Materials, 4[2] 2017, 021030.

[237] Wang, J., Sui, X., Shi, W., Pan, J., Zhang, S., Liu, F., Wei, S.H., Yan, Q., Huang, B., Physical Review Letters, 119[25] 2017, 256402. https://doi.org/10.1103/PhysRevLett.119.256402

[238] Yan, M., Huang, H., Zhang, K., Wang, E., Yao, W., Deng, K., Wan, G., Zhang, H., Arita, M., Yang, H., Sun, Z., Yao, H., Wu, Y., Fan, S., Duan, W., Zhou, S., Nature Communications, 8[1] 2017, 257. https://doi.org/10.1038/s41467-017-00280-6

[239] Pavlosiuk, O., Kaczorowski, D., Scientific Reports, 8[1] 2018, 11297. https://doi.org/10.1038/s41598-018-29545-w

[240] Ma, H., Chen, P., Li, B., Li, J., Ai, R., Zhang, Z., Sun, G., Yao, K., Lin, Z., Zhao, B., Wu, R., Tang, X., Duan, X., Duan, X., Nano Letters, 18[6] 2018, 3523-3529. https://doi.org/10.1021/acs.nanolett.8b00583

[241] Xiao, R.C., Cheung, C.H., Gong, P.L., Lu, W.J., Si, J.G., Sun, Y.P., Journal of Physics - Condensed Matter, 30[24] 2018, 245502. https://doi.org/10.1088/1361-

648X/aac298

[242] Kunduru, L., Roshan, S.C.R., Yedukondalu, N., Sainath, M., AIP Conference Proceedings, 1966, 2018, 020029.

[243] Luo, X., Chen, F.C., Pei, Q.L., Gao, J.J., Yan, J., Lu, W.J., Tong, P., Han, Y.Y., Song, W.H., Sun, Y.P., Applied Physics Letters, 110[9] 2017, 092401. https://doi.org/10.1063/1.4977708

[244] Koepernik, K., Kasinathan, D., Efremov, D.V., Khim, S., Borisenko, S., Büchner, B., Van Den Brink, J., Physical Review B, 93[20] 2016, 201101. https://doi.org/10.1103/PhysRevB.93.201101

[245] Soluyanov, A.A., Gresch, D., Wang, Z., Wu, Q., Troyer, M., Dai, X., Bernevig, B.A., Nature, 527[7579] 2015, 495-498. https://doi.org/10.1038/nature15768

[246] Huang, C., Narayan, A., Zhang, E., Liu, Y., Yan, X., Wang, J., Zhang, C., Wang, W., Zhou, T., Yi, C., Liu, S., Ling, J., Zhang, H., Liu, R., Sankar, R., Chou, F., Wang, Y., Shi, Y., Law, K.T., Sanvito, S., Zhou, P., Han, Z., Xiu, F., ACS Nano, 12[7] 2018, 7185-7196. https://doi.org/10.1021/acsnano.8b03102

[247] Weng, H., Fang, C., Fang, Z., Dai, X., Physical Review B, 94[16] 2016, 165201. https://doi.org/10.1103/PhysRevB.94.165201

[248] Guo, S.D., Wang, Y.H., Lu, W.L., New Journal of Physics, 19[11] 2017, 113044. https://doi.org/10.1088/1367-2630/aa96f7

[249] Lu, J., Zheng, G., Zhu, X., Ning, W., Zhang, H., Yang, J., Du, H., Yang, K., Lu, H., Zhang, Y., Tian, M., Physical Review B, 95[12] 2017, 125135. https://doi.org/10.1103/PhysRevB.95.125135

[250] Hu, J., Zhu, Y., Gui, X., Graf, D., Tang, Z., Xie, W., Mao, Z., Physical Review B, 97[15] 2018, 155101. https://doi.org/10.1103/PhysRevB.97.155101

[251] Liu, G., Jin, L., Dai, X., Chen, G., Zhang, X., Physical Review B, 98[7] 2018, 075157. https://doi.org/10.1103/PhysRevB.98.075157

[252] Chang, G., Xu, S.Y., Zhou, X., Huang, S.M., Singh, B., Wang, B., Belopolski, I., Yin, J., Zhang, S., Bansil, A., Lin, H., Hasan, M.Z., Physical Review Letters, 119[15] 2017, 156401. https://doi.org/10.1103/PhysRevLett.119.156401

[253] Lan, H.S., Chang, S.T., Liu, C.W., Physical Review B, 95[20] 2017, 201201. https://doi.org/10.1103/PhysRevB.95.201201

[254] Kim, M., Wang, C.Z., Ho, K.M., Physical Review B, 96[20] 2017, 205107. https://doi.org/10.1103/PhysRevB.96.205107

[255] Luo, X., Xiao, R.C., Chen, F.C., Yan, J., Pei, Q.L., Sun, Y., Lu, W.J., Tong, P., Sheng, Z.G., Zhu, X.B., Song, W.H., Sun, Y.P., Physical Review B, 97[20] 2018, 205132. https://doi.org/10.1103/PhysRevB.97.205132

[256] Guo, P.J., Yang, H.C., Liu, K., Lu, Z.Y., Physical Review B, 98[4] 2018, 045134. https://doi.org/10.1103/PhysRevB.98.045134

[257] Scholz, M.R., Rogalev, V.A., Dudy, L., Reis, F., Adler, F., Aulbach, J., Collins-Mcintyre, L.J., Duffy, L.B., Yang, H.F., Chen, Y.L., Hesjedal, T., Liu, Z.K., Hoesch, M., Muff, S., Dil, J.H., Schäfer, J., Claessen, R., Physical Review B, 97[7] 2018, 075101. https://doi.org/10.1103/PhysRevB.97.075101

[258] Kharitonov, M., Mayer, J.B., Hankiewicz, E.M., Physical Review Letters, 119[26] 2017, 266402. https://doi.org/10.1103/PhysRevLett.119.266402

[259] Gao, H., Kim, Y., Venderbos, J.W.F., Kane, C.L., Mele, E.J., Rappe, A.M., Ren, W., Physical Review Letters, 121[10] 2018, 106404. https://doi.org/10.1103/PhysRevLett.121.106404

[260] Enderlein, C., Ramos, S.M., Bittencourt, M., Continentino, M.A., Brewer, W., Baggio-Saitovich, E., Journal of Applied Physics, 114[14] 2013, 143711. https://doi.org/10.1063/1.4825073

[261] Kim, Y., Wieder, B.J., Kane, C.L., Rappe, A.M., Physical Review Letters, 115[3] 2015, 036806. https://doi.org/10.1103/PhysRevLett.115.036806

[262] Kim, J., Kim, H.S., Vanderbilt, D., Physical Review B, 98[15] 2018, 155122. https://doi.org/10.1103/PhysRevB.98.155122

[263] Ebrahimian, A., Dadsetani, M., Frontiers of Physics, 13[5] 2018, 137309. https://doi.org/10.1007/s11467-018-0815-x

[264] Huang, H., Jin, K.H., Liu, F., Physical Review Letters, 120[13] 2018, 136403. https://doi.org/10.1103/PhysRevLett.120.136403

[265] Weng, H., Fang, C., Fang, Z., Dai, X., Physical Review B, 93[24] 2016, 241202. https://doi.org/10.1103/PhysRevB.93.241202

[266] Guo, S.D., Liu, B.G., Journal of Physics - Condensed Matter, 30[10] 2018, 105701. https://doi.org/10.1088/1361-648X/aaab32

[267] Huang, H., Jiang, W., Jin, K.H., Liu, F., Physical Review B, 98[4] 2018, 045131. https://doi.org/10.1103/PhysRevB.98.045131

[268] Lepori, L., Fulga, I.C., Trombettoni, A., Burrello, M., Physical Review A, 94[5] 2016, 053633. https://doi.org/10.1103/PhysRevA.94.053633

[269] Schaffer, R., Lee, E.K.H., Lu, Y.M., Kim, Y.B., Physical Review Letters, 114[11] 2015, 116803. https://doi.org/10.1103/PhysRevLett.114.116803

[270] Fu, B., Fan, X., Ma, D., Liu, C.C., Yao, Y., Physical Review B, 98[7] 2018, 075146. https://doi.org/10.1103/PhysRevB.98.075146

[271] Wang, R., Jin, Y.J., Zhao, J.Z., Chen, Z.J., Zhao, Y.J., Xu, H., Physical Review B, 97[19] 2018, 195157. https://doi.org/10.1103/PhysRevB.97.195157

[272] Li, S., Liu, Y., Fu, B., Yu, Z.M., Yang, S.A., Yao, Y., Physical Review B, 97[24] 2018, 245148. https://doi.org/10.1103/PhysRevB.97.245148

[273] Fujita, T.C., Uchida, M., Kozuka, Y., Ogawa, S., Tsukazaki, A., Arima, T., Kawasaki, M., Applied Physics Letters, 108[2] 2016, 022402. https://doi.org/10.1063/1.4939742

[274] Oh, T., Ishizuka, H., Yang, B.J., Physical Review B, 98[14] 2018, 144409. https://doi.org/10.1103/PhysRevB.98.144409

[275] Chen, T., Shao, D., Lu, P., Wang, X., Wu, J., Sun, J., Xing, D., Physical Review B, 98[14] 2018, 144105. https://doi.org/10.1103/PhysRevB.98.144105

[276] Kasinathan, D., Koepernik, K., Tjeng, L.H., Haverkort, M.W., Physical Review B, 91[19] 2015, 195127. https://doi.org/10.1103/PhysRevB.91.195127

[277] Lado, J.L., Pardo, V., Baldomir, D., Physical Review B, 88[15] 2013, 155119. https://doi.org/10.1103/PhysRevB.88.155119

[278] Gordon, J.S., Kee, H.Y., Physical Review B, 97[19] 2018, 195106. https://doi.org/10.1103/PhysRevB.97.195106

[279] Fang, C., Lu, L., Liu, J., Fu, L., Nature Physics, 12[10] 2016, 936-941.

[280] Wan, X., Turner, A.M., Vishwanath, A., Savrasov, S.Y., Physical Review B, 83[20] 2011, 205101. https://doi.org/10.1103/PhysRevB.83.205101

[281] Craco, L., Pereira, T.A.D.S., Ferreira, S.R., Carara, S.S., Leoni, S., Physical Review B, 98[3] 2018, 035114. https://doi.org/10.1103/PhysRevB.98.035114

[282] Sun, Z., Xiang, Z., Wang, Z., Zhang, J., Ma, L., Wang, N., Shang, C., Meng, F., Zou, L., Zhang, Y., Chen, X., Science Bulletin, 63[23] 2018, 1539-1544. https://doi.org/10.1016/j.scib.2018.11.006

[283] Goh, W.F., Pickett, W.E., Physical Review B, 98[12] 2018, 125147. https://doi.org/10.1103/PhysRevB.98.125147

[284] Goh, W.F., Pickett, W.E., Physical Review B, 97[3] 2018, 035202.

https://doi.org/10.1103/PhysRevB.97.035202

[285] Yao, H., Zhu, M., Jiang, L., Zheng, Y., Journal of Physics - Condensed Matter, 30[28] 2018, 285501. https://doi.org/10.1088/1361-648X/aac793

[286] Chan, Y.H., Chiu, C.K., Chou, M.Y., Schnyder, A.P., Physical Review B, 93[20] 2016, 205132. https://doi.org/10.1103/PhysRevB.93.205132

[287] Chen, W., Luo, K., Li, L., Zilberberg, O., Physical Review Letters, 121[16] 2018, 166802. https://doi.org/10.1103/PhysRevLett.121.166802

[288] Lv, B.Q., Feng, Z.L., Xu, Q.N., Gao, X., Ma, J.Z., Kong, L.Y., Richard, P., Huang, Y.B., Strocov, V.N., Fang, C., Weng, H.M., Shi, Y.G., Qian, T., Ding, H., Nature, 546[7660] 2017, 627-631. https://doi.org/10.1038/nature22390

[289] Guo, S.D., Journal of Physics - Condensed Matter, 29[43] 2017, 435704. https://doi.org/10.1088/1361-648X/aa8939

[290] Chi, Z., Chen, X., An, C., Yang, L., Zhao, J., Feng, Z., Zhou, Y., Zhou, Y., Gu, C., Zhang, B., Yuan, Y., Kenney-Benson, C., Yang, W., Wu, G., Wan, X., Shi, Y., Yang, X., Yang, Z., npj Quantum Materials, 3[1] 2018, 102-109.

[291] Li, G., Sun, Y., Rao, J., Wu, J., Kumar, A., Xu, Q.N., Fu, C., Liu, E., Blake, G.R., Werner, P., Shao, B., Liu, K., Parkin, S., Liu, X., Fahlman, M., Liou, S.C., Auffermann, G., Zhang, J., Felser, C., Feng, X., Advanced Energy Materials, 8[24] 2018, 1801258. https://doi.org/10.1002/aenm.201801258

[292] Kumar, N., Sun, Y., Xu, N., Manna, K., Yao, M., Süss, V., Leermakers, I., Young, O., Förster, T., Schmidt, M., Borrmann, H., Yan, B., Zeitler, U., Shi, M., Felser, C., Shekhar, C., Nature Communications, 8[1] 2017, 1642-1650. https://doi.org/10.1038/s41467-017-01758-z

[293] Li, G., Sun, Y., Rao, J., Wu, J., Kumar, A., Xu, Q.N., Fu, C., Liu, E., Blake, G.R., Werner, P., Shao, B., Liu, K., Parkin, S., Liu, X., Fahlman, M., Liou, S.C., Auffermann, G., Zhang, J., Felser, C., Feng, X., Advanced Energy Materials, 8[24] 2018, 1801258. https://doi.org/10.1002/aenm.201801258

[294] Shekhar, C., Nayak, A.K., Sun, Y., Schmidt, M., Nicklas, M., Leermakers, I., Zeitler, U., Skourski, Y., Wosnitza, J., Liu, Z., Chen, Y., Schnelle, W., Borrmann, H., Grin, Y., Felser, C., Yan, B., Nature Physics, 11[8] 2015, 645-649.

[295] Fu, C., Guin, S.N., Watzman, S.J., Li, G., Liu, E., Kumar, N., Süß, V., Schnelle, W., Auffermann, G., Shekhar, C., Sun, Y., Gooth, J., Felser, C., Energy and Environmental Science, 11[10] 2018, 2813-2820.

https://doi.org/10.1039/C8EE02077A

[296] Wang, Z., Zheng, Y., Shen, Z., Lu, Y., Fang, H., Sheng, F., Zhou, Y., Yang, X., Li, Y., Feng, C., Xu, Z.A., Physical Review B, 93[12] 2016, 121112. https://doi.org/10.1103/PhysRevB.93.121112

[297] Xu, N., Weng, H.M., Lv, B.Q., Matt, C.E., Park, J., Bisti, F., Strocov, V.N., Gawryluk, D., Pomjakushina, E., Conder, K., Plumb, N.C., Radovic, M., Autès, G., Yazyev, O.V., Fang, Z., Dai, X., Qian, T., Mesot, J., Ding, H., Shi, M., Nature Communications, 7, 2016, 11006. https://doi.org/10.1038/ncomms11006

[298] Cheng, W., Fu, Y.L., Ying, M.J., Zhang, F.S., Chinese Physics Letters, 34[12] 2017, 127101. https://doi.org/10.1088/0256-307X/34/12/127101

[299] Leahy, I.A., Lin, Y.P., Siegfried, P.E., Treglia, A.C., Song, J.C.W., Nandkishore, R.M., Lee, M., Proceedings of the National Academy of Sciences of the United States of America, 115[42] 2018, 10570-10575. https://doi.org/10.1073/pnas.1808747115

[300] Razzoli, E., Zwartsenberg, B., Michiardi, M., Boschini, F., Day, R.P., Elfimov, I.S., Denlinger, J.D., Süss, V., Felser, C., Damascelli, A., Physical Review B, 97[20] 2018, 201103. https://doi.org/10.1103/PhysRevB.97.201103

[301] Takahashi, R., Hirayama, M., Murakami, S., Physical Review B, 96[15] 2017, 155206. https://doi.org/10.1103/PhysRevB.96.155206

[302] Zhu, Z., Li, M., Li, J., Physical Review B, 94[15] 2016, 155121. https://doi.org/10.1103/PhysRevB.94.155121

[303] Pshenay-Severin, D.A., Ivanov, Y.V., Burkov, A.T., Journal of Physics - Condensed Matter, 30[47] 2018, 475501. https://doi.org/10.1088/1361-648X/aae6d1

[304] Chang, G., Xu, S.Y., Zheng, H., Singh, B., Hsu, C.H., Bian, G., Alidoust, N., Belopolski, I., Sanchez, D.S., Zhang, S., Lin, H., Hasan, M.Z., Scientific Reports, 6, 2016, 38839. https://doi.org/10.1038/srep38839

[305] Chang, G., Xu, S.Y., Wieder, B.J., Sanchez, D.S., Huang, S.M., Belopolski, I., Chang, T.R., Zhang, S., Bansil, A., Lin, H., Hasan, M.Z., Physical Review Letters, 119[20] 2017, 206401. https://doi.org/10.1103/PhysRevLett.119.206401

[306] Singh, B., Chang, G., Chang, T.R., Huang, S.M., Su, C., Lin, M.C., Lin, H., Bansil, A., Scientific Reports, 8[1] 2018, 10540. https://doi.org/10.1038/s41598-018-28644-y

[307] Nie, S., Xu, G., Prinz, F.B., Zhang, S.C., Proceedings of the National Academy of Sciences of the United States of America, 114[40] 2017, 10596-10600. https://doi.org/10.1073/pnas.1713261114

Topological Semimetals
Materials Research Foundations **48** (2019)

Materials Research Forum LLC
doi: http://dx.doi.org/10.21741/9781644900154

Keyword Index

www.ingramcontent.com/pod-product-compliance
Lightning Source LLC
Chambersburg PA
CBHW071645210326
41597CB00017B/2125